W9-BKA-440

Dictionary of

Electronic Packaging,

Microelectronic, and

Interconnection Terms

Dictionary of Electronic Packaging, Microelectronic, and Interconnection Terms

by Martin B. Miller

Technology Seminars Inc.

P.O. Box 487

Lutherville, Maryland 21093

Table of Contents

INTERCONNECTIONS–
from the Chip
to the Backplane
and Beyond...

- ❏ R/flex® TAB Materials
- ❏ Multimetal Layer TAB Circuits
- ❏ R/flex® Flexible Circuit Materials
- ❏ Mektron® Flexible Circuits
- ❏ Smartflex® Flexible Circuit Assemblies
- ❏ BEND/flex® Formable Laminates
- ❏ ROHSI® Rogers High Speed Interconnection Systems

- ❏ RO2800™ Low Dielectric Constant Circuit Material
- ❏ INVISICON® and ISOCON™ Connection Systems
- ❏ Mektron® Power Bus Bars
- ❏ Micro/Q® Decoupling Capacitors
- ❏ Q-PAC® and Magna/PAC® Power Distribution Capacitors

Technology for tomorrow built on TQC today

Call or Write for Details.

 **ROGERS**

Rogers Corporation
One Technology Drive
Rogers, Connecticut 06263
203 774-9605
FAX 203 774-9630

NEW-GENERATION SOLDER PASTE

INTEGRATION OF PASTE TECHNOLOGY & POWDER TECHNOLOGY

New products, Service, and Quality

Service: Customize and personalize our service to your needs.

Quality: Your commitment to production yield and product reliability is our commitment.

New Product: Design products for application-specific.

Finer pitch, no-clean, water-cleaning, or your ideal product . . . .

WORK TOGETHER AS PARTNERS

INTERNATIONAL ELECTRONIC MATERIALS CORP.

a subsidiary of Advanced Metals Technology, Inc.   30275 Bainbridge Road   Building B-8   Cleveland, Ohio 44139
Telephone (216) 349-1960   Fax (216) 349-2206

Dow Corning Presents

A Design Digest of Electronic Reliability

Designing a high-performance electronic product is only half the battle. Protecting it is the other half. Demand our high-reliability protective materials and you have won. Here is what we mean:

New thermally conductive elastomer conducive many uses.

or thermal dissipation that protects erformance, choose new SYLGARD® 3-6605 thermally conductive elas- mer. It is especially useful as a ther- ally conductive adhesive to bond ramic hybrid circuit packages to at sinks. Other uses include bond- g or potting power supplies, coils, avy-duty relays, and similar units. imerless, too.

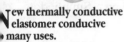

rotective silicone gels lieve stress and nprove reliability.

hen you encapsulate omponents in a licone gel, you protect em against hostile vironments, temperatures low as -80°C and more. OW CORNING® silicone gels so cushion components against ress and vibration. We offer a broad ioice of one component and two omponent materials, for room tem- erature and heat cure processes.

Coating makes repairs on PC boards easier.

A PC board coating simply has to be reliable. And with DOW CORNING 1-2577 conformal coating you get more: easy repairability when compo- nents need replacing. Add superior protection in extreme environments, versatile processing with dipping or spraying, and you have all the reasons major automotive and military electronics companies choose
DOW CORNING 1-2577.

Stuck for the right adhesive?

Announcing new additions to the big family of Dow Corning adhesives for electronics. First, a one-part primerless silicone adhesive with rapid heat cure. And, a fast, room temperature cure two-part system. Also new – fast, tack- free, primerless, neutral cure RTV's. There's no adhesive problem too sticky for Dow Corning.

There's More.

There's also our Electronics Industry Team whose mission is to put Dow Corning's full line of coatings, adhesives, sealants, and encapsulants to work on your specific applications.

DOW CORNING®

Dow Corning and Sylgard are registered trademarks of Dow Corning Corporation.

1989 Dow Corning Corporation.

TECHNICAL BULLETIN
ELECTRONIC MATERIALS

THE FOLLOWING PRODUCTS ARE AVAILABLE FOR A WIDE RANGE OF *HYBRID CIRCUIT* AND *SEMICONDUCTOR* APPLICATIONS IN THE ASSEMBLY PROCESSES.

Electronic Grade Materials

Clad Metal Parts--MP/CMX-100--Available in a wide range of metals, fabricated to customer specifications.

Die Attach Preforms--MP/DAX-200--For attaching semiconductor die into packages.

Evaporation Sources--MP/ESX-300--For semiconductor and thin film processing. Available in high purity Aluminum and Aluminum Alloys. Also available in Low Alpha Grade. Supplied as charges, slugs, or wire.

Solders--MP/SOX-700--Available in various Low Oxide metals and alloys. Supplied as ingots or preforms.

Solder Creams--MP/SCX-710--Available in various Low Oxide metals and alloys. Four viscosity ranges and six fluxes are available to accommodate differing process techniques.

Sputtering Targets--MP/STX-800--For semiconductor and thin film processing. Available in high purity Aluminum and Aluminum Alloys. Also available in Low Alpha Grade.

Small Equipment, Supplies, and Tools

Lid Locating Fixtures--AP/LLX-100--For aligning the lid and package seal ring prior to applying the sealing clip. Available in a range of standard sizes or can be custom fabricated.

Sealing Clips--AP/SCX-200--For holding the lid and ceramic package together during the furnace sealing cycle. Available in a range of standard sizes or can be custom fabricated.

Sealing Clip Pressure Gauge--AP/PGX-300--For testing individual sealing clip pressure.

Electro Assembly Source, In

7 Sundbury Drive, New City, New York 10956 · Phone: (914) 634-1327 · Fax: (914) 634-09

Preface

Progress in electronic packaging, microcroelectronics, and interconnection technologies has been, and continues to be, rapid. Indeed, it is increasingly recognized that this technology area is the critical limiting factor to the success of modern high performance electronic systems. Yet, progress is often impeded by the lack of understanding of the language in this field. There is, in fact, a language barrier. This is due to the multidisciplinary nature of electronic packaging and interconnecting. Most of those involved in this field are specialists - mechanical, electrical, materials, engineering, manufacturing, marketing, and others. Specialists in one discipline are not usually very familiar with the other discipline. A major objective of this book is to bridge this barrier by including the important terms in all of these disciplines. To further assist the reader, a thorough listing of often confusing acronyms is included.

This book will provide a consistent and ready reference for the desks of all those involved in any discipline or aspect of electronic packaging, microelectronics or interconnections. The text includes approximately twenty-five hundred multidiscipline terms and definitions. Along with definitions, acronyms are included in brackets where acronyms apply. Then, following the terms and definitions part of the text, the acronyms and their meanings are listed, again in easy to use alphabetical order. Comments and suggestions are much welcome, including recommendations for new terms in future printings and editions of this book.

No book such as this can ever be the work of one person. Hence, I would like to express my deep appreciation to all of my friends and associates who have been so helpful in the preparation of this book. Their assistance and advice have been invaluable, both during preparation and review of the text.

Martin B. Miller

About the Author:

Mr. Miller worked for the Westinghouse Electric Corporation for more than thirty years where he was deeply involved in materials engineering as well as mechanical engineering before retiring in 1989. He managed several engineering development laboratories where he was involved in both the design and development of electronic packaging as well as the manufacture of a wide variety of electronic packages and assemblies. He is a graduate of Loyola College and has authored numerous articles and has been awarded several patent disclosures in the field of electronic packaging, microelectronic, and interconnection, including the electrical interconnection of multilayer substrates and thermal management of surface mount devices.

Ablative Material A material which dissipates heat by vaporizing or melting.

Abrasion Stripper A mechanical device equipped with buffing wheels which scrape insulation from conductors. (Also known as Buffing Stripper).

Abrasion Tester A laboratory testing machine which determines the abrasive resistance of the insulation on wire and cable.

Abrasive Trimming Adjusting film resistors to nominal valves by notching the resistor with a stream of abrasive material such as aluminum oxide.

Absorption (1) The retention of moisture by a substance. (2) The dissipation or transformation of energy as it passes through a medium.

AC A circuit or voltage source in which voltage changes from a negative to a positive polarity as a sinusoidal function of time.

Accelerated Aging To rapidly induce deterioration of a material, system or device in a short time by increasing voltage, temperature, etc. above normal operating values.

Accelerated Stress Test A test which is conducted at a higher stress level than in normal operation and at a shorter time for the purpose of producing a failure.

Acceleration The rate of change of velocity.

Accelerator A chemical used to speed up a reaction or cure. The term accelerator is often used interchangeably with the term promoter. An accelerator is often used along with a catalyst, hardener, or curing agent.

Access Hole A hole which is drilled through successive layers of a multilayer printed wiring board to gain access to an area on one of the inner layers.

Accessories An assortment of mechanical parts such as clamps and other such hardware which make up the complete connector assembly and are attached to the connector.

Accordion A "Z" shaped spring contact. It is shaped in the form of a Z to give the spring a high deflection without excessive stress. These contacts are used in some printed circuit connectors.

Acetal A rigid thermoplastic material which can be molded or extruded. They are noted for their high tensile and flexural strength, good solvent and moisture resistance.

Acrylic A synthetic resin which is made from acrylic acid or a derivative thereof. Offers good flame resistance and has excellent transparency properties.

Acrylic Modifier Compounds which are added to acrylic resins to change the physical properties.

Acrylic Resin A thermosetting, transparent resin.

Acrylonitrile-Butadiene-Styrene (ABS) A low cost thermoplastic which exhibits dimensional stability, high impact resistance, excellent mechanical properties and high surface hardness over a wide temperature range.

Activating A chemical treatment which conditions the surface of nonconductive materials for improved adhesion of electroless deposited materials.

Active Area The area of an electronic package where the substrate is attached.

Active Components Electronic devices such as transistors, diodes, etc. which are capable of controlling voltages and currents in producing gain or switching actions in circuits. (Same as Active Element.)

Active Element See Active Component.

Active Network A network containing both active and passive elements.

Active Trimming Adjusting an electrical circuit with the power "on" to achieve a required functional output. Usually involves the trimming of thick film resistors.

Actuator A part of a switch which, when activated by an external force, causes the switch to open or close.

Additive Process A process for obtaining conductive patterns by the selective deposition of conductive material on an unclad base material.

Add-On-Component Any type of component that is attached to a circuit to complete the circuit function. (Same as Add-On-Device)

Add-On Device See Add-On-Component.

Adhere To cause two surfaces to be held together by adhesion.

Adherend A body which is held to another body by an adhesive.

Adhesion The state in which two surfaces are held together by interfacial forces which may consist of valence forces or interlocking action, or both.

Adhesion Promotion A chemical process which conditions the surface of a plastic or plastic parts and provides excellent adhesion of subsequent metal depositions.

Adhesion Promotor A chemical which prepares the bonding surfaces to improve the bond strength. (Also called a Primer)

Adhesive A substance capable of holding materials together by surface attachment.

Adhesive, Contact An adhesive that is apparently dry to the touch and which will adhere to itself instantaneously upon contact; also called contact-bond adhesive or dry-bond adhesive.

Adhesive, Heat-Activated A dry adhesive film that is rendered tacky or fluid by application of heat or heat and pressure to the assembly.

Adhesive, Pressure-Sensitive A viscoelastic material which in solvent-free form remains permanently tacky. Such material will adhere instantaneously to most solid surfaces with the application of very slight pressure.

Adhesive, Room-Temperature-Setting An adhesive that sets in the temperature range of 20 to 30 degree C (68 to 86 degree F), in accordance with the limits for Standard Room Temperature specified in ASTM Methods D 618, Conditioning Plastics and Electrical Insulating Materials for Testing.

Adhesive, Solvent An adhesive having a volatile organic liquid as a vehicle.

Adhesive, Solvent Activated A dry adhesive film that is rendered tacky just prior to use by application of a solvent.

Adjacent Conductor A conductor located next to another conductor.

Admittance The unrestricted flow of alternating current in a circuit. It is the reciprocal of impedance.

Adsorption The adhesion of gas or liquid molecules to the surface of solids or liquids with which they are in contact.

Advanced Statistical Analysis Program (ASTAP) A simulated circuit analysis program which performs DC, time domain, and frequency domain simulations to predict operating tolerances by statistical analysis. Included is a transmission line analysis program.

Aging The change in properties of a material with time under specific conditions such as varying the temperature, pressure, humidity, etc.

Aging, Accelerated The change in properties of a material under accelerated conditions, such as elevated temperatures, humidity, electrical or mechanical stresses, etc.

Aging, Natural Factors which effect a material under normal environmental conditions such as oxidation, reaction with sulphur and other chemicals, formation of reaction layers between base material and coating materials.

Aging, Steam An artificial, accelerated treatment to the outer surfaces of components and boards to determine whether they can withstand storage requirements.

Agitation A process involving motion of a medium such as shaking, stirring or vibration of a liquid.

Air Gap The space between conductors, lands, pads, etc., on a PWB.

Alcohols Characterized by the fact that they contain the hydroxyl (-OH) group, they are valuable starting points for the manufacture of synthetic resins, synthetic rubbers, and plasticizers.

Aldehydes In general, aldehydes are volatile liquids with sharp, penetrating odors and are slightly less soluble in water than corresponding alcohols. The group (-CHO), which characterizes all aldehydes, contains the most active form of the carbonyl radical and makes the aldehydes important as organic synthetic agents. They are widely used in industry as chemical building blocks in organic synthesis.

Align To accurately position or locate circuits, images, patterns of a photomask over each other or over etched or screened patterns.

Aliphatic Hydrocarbon See Hydrocarbon.

Alkyd A thermosetting resin which has excellent electrical properties and used for molding a wide variety of electrical components.

Alligator Clip A mechanical device shaped like the jaws of an alligator. Used as a temporary connection at the end of test leads.

Alloy A metal formed by combining two or more other metals.

Alloy Junction A point at which one or more impurity metals are melted on a semi conductor wafer to form p or n regions.

Alpha Particle A small electrically charged particle which is thrown off at very high speeds by radioactive materials and capable of generating electron/hole pairs in microelectronic devices causing soft errors in some components.

Alumina The polycrystalline form aluminum oxide, Al_2O_3. A high quality ceramic compound used to make ceramic substrates and chip carriers.

Aluminum A lightweight metal with good corrosion resistance, excellent electrical and thermal conductivity, and used extensively in electrical and aircraft construction.

Aluminum-Steel Conductor A composite of aluminum and steel wires where the aluminum wires surround the steel wires.

Ambient Temperature The temperature of the surrounding cooling medium such as gas or liquid, which comes into contact with the heated parts of the apparatus.

Amine Adduct Products of the reaction of an amine with a deficiency of a substance containing epoxy groups.

Amino Plastics A group of thermosetting plastics which include melamine and urea used in molding compounds.

Analog Circuits Circuits which provide a continuous relationship between input and output.

Analog Computer A device which measures voltages, linear lengths, resistances, light intensities, and can be manipulated by the computer.

Analog-To-Digital-Converter A device which converts an analog signal into a proportional digital signal.

Angle Of Attack In screen printing, it is the angle between the face of the squeegee and surface of the screen. (Also called Attack Angle.)

Angle Connector A type of connector which joins two conductors, end to end, at a specific angle.

Angled Bond Two bonds which are not in a straight line.

Angstrom Approximately 4×10^{-9} inches or 10^{-8} centimeters or 10^{-4} microns.

Anhydrides Anhydrides result from the removal of water from organic acids. In general, they are liquid that boil at higher temperatures than the corresponding acids. They are particularly valuable for reactions with the alcohol groups found in cellulose, phenols, sugars, and vegetable oils, yielding esters which are less water-soluble than the original alcohol.

Annular Conductor A number of stranded wires that are twisted in three reversed concentric layers around a core.

Anneal To heat and cool at a slow or prescribed rate to relieve mechanical stresses.

Annular Ring The circular strip of conductive material that completely surrounds a hole.

Anode The positive (+) terminal in electroplating toward which the negative (-) ions flow in the plating baths. The part to be plated is usually the anode.

Anodic Silver A bar of silver used in electroplating baths. The silver bar is the sacrificial anode during the plating process.

Anodize   A proces whereby an insulating oxide layer is formed or deposited on a metal part by electrolytic oxidation.

Anodization   An electrochemical oxidation process used to change the value of thin film resistors.

Antioxidant   A chemical used in the formulation of plastics to prevent or slow down the oxidation of material exposed to the air.

Anti-Rotation   A connector design with locking provisions to control positive orientation.

Antistatic Agents   Agents which, when added to the molding material or applied onto the surface of the molded object, make it less conductive (thus hindering the fixation of foreign particles).

Anvil   That part of a crimping die which positions and supports the terminal during crimping.

Application Specific Integrated Circuit (ASIC)
An IC chip designed for a specific application or product.

Aramid   A generic description defining a group of high temperature aromatic polyamides. DuPont Kevlar is a high modulus fiber form aramid.

Arc-Over Voltage   The minimum voltage required to cause an arc between electrodes separated by a gas or liquid insulation.

Area Array Tab   An array of pads which are located along the edge as well as the inner surface area of a substrate. It is practical on complex dies and IC's where peripheral pad pitch cannot be reduced and more I/O's must be accommodated.

Arc Resistance   The time required for an arc to establish a conductive path in a material.

Armored Cable   An electrical cable equipped with a metal wrapping usually for mechanical protection.

Aromatic   An organic molecule built around the hexagonal six carbon benzene structure. Usually more thermally stable than linear or branched aliphatic structures.

Aromatic Hydrocarbon   See Hydrocarbon.

Array   A group of elements or circuits arranged in rows and columns on a substrate or printed wiring board.

Array Device   Devices without individual enclosures, each one having at least one of its electrodes connected to a common conductor or all connected in series.

Artwork An accurately scaled configuration produced at an enlarged ratio to enable the product to be made therefrom by photographic reduction. Used to create screens and masks.

As-Fired (1) The smoothness of ceramic substrates. (2) The values of thick film resistors as they emerge from the furnace before polishing or trimming.

Aspect Ratio (1) The ratio of the length of a film resistor to its width (L/W); equal to the number of squares of the resistor. (2) the ratio of the length or depth of a hole to its diameter (unplated) in a printed wiring board.

Assembly A group of materials or parts, including adhesive, which has been placed together for bonding or which has been bonded together.

Assembly Drawing A drawing of all the parts and subassemblies mounted in their proper location and orientation.

Assembly/Rework This term refers to the processes used in the initial attachment and interconnection of a device to a package and also the rework in the replacement of a device which includes the removal of the device, preparation of the pad or area for the new component, mounting, and interconnections. This is sometimes required for either replacing a defective component or incorporating an engineering change.

A-Stage The initial uncured stage in the curing of a thermosetting resin. In this stage, resins can be heated and caused to flow, thereby allowing final curing into the desired shape. The term A stage is used to describe the uncured stage in the curing reaction, and the term C stage is sometimes used to describe the cured resin.

Atactic When the radical groups are arranged heterogeneously around the carbon chain.

Attack Angle See Angle of Attack.

Attenuation (1) The decrease in amplitude of a signal from one point to another in an electrical system. (2) The power loss in a length of cable. The loss is expressed in decibels. (3) In fiber optics it is the lessening of light intensity due to absorption and scattering of the light as it travels through the optic fibers.

Au/Ge An alloy of gold and germanium usually in a 88/12 ratio. It is used in higher temperature substrate and die attachment.

Autoclave A closed vessel for conducting a chemical reaction or other operations under pressure and heat.

Autoclave Molding After lay-up, the entire assembly is placed in a steam autoclave at 50 to 100 lb/in^2. Additional pressure achieves higher reinforcement loadings and improved removal of air.

Au/Sn A solder alloy of gold/tin usually in a 80/20 ratio used for sealing gold plated packages.

Automated A combination of mechanical and electrical techniques and facilities programmed to perform a function without human effort.

Automated Wire Bonding A computer aided high production process in which wire bonding is performed by either thermocompression or ultrasonic methods.

Automatic Dip Insertion A dual in line package which is machine inserted in its proper placement. The package height is controlled at a predetermined distance above the printed wiring board by shoulders on the leads.

Axial Displacement An incremental difference between the initial position and a final position after a force is applied along the components axis.

Axial Lead A wire coming out from the end of a discrete device; along the axis of the device rather than from the sides.

Axial Strain Young's Modulus of Elasticity indicating strain applied along the axis of a material.

Azeotrope A mixture of several polar and nonpolar solvents to remove both polar and non polar contaminates. The mixture has a constant boiling point and cannot be separated by distillation.

Back Bonding Bonding IC's to a substrate by applying adhesive to the back side of the chips. (Also called Back Mounting.)

Back End Of The Line (BEOL) Integrated circuit fabrication processes such as: (1) Dicing of the wafer into individual IC's (2) Wire bonding of the transistors, resistors, etc. to the pads on the substrate, and chip-to-package connections.

Backfill After evacuating an electronic package, to fill the package with a dry inert gas prior to hermetic sealing.

Back Mounted A connector that is mounted from inside of a box/panel and its mounting flanges located inside the equipment.

Back Mounting Same as Back Bonding.

Backplane Panel An interconnection panel in which PWA or integrated circuit packages can be plugged or mounted.

B

Back Radius The radius of the trailing edge of a bonding tool foot.

Backshell Mold A mold that is used for molding the back of a connector after the cable connections have been installed.

Backside Metallurgy (BSM) The brazing of pins to metallized pads, which are connected to internal conductors within a multilayer ceramic package.

Bag Molding A method of applying pressure during bonding or molding in which a flexible cover, usually in connection with a rigid die or mold, exerts pressure on the material being molded, through the application of air pressure or drawing of a vacuum.

Bake-Out An elevated temperature cycle which expels unwanted gases and moisture prior to sealing the package.

Ball Bond A type of wire bond in which a interconnection wire that was flame cut to produce a ball shaped end is deformed by thermocompression tool on a metalized pad. Also called a nail head bond because of its flattened appearance.

Ball Limiting Metallurgy (BLM) The metallurgy which controls the size and area of solder connections and limits the flow of solder balls to the desired area and provides adhesion and contact to the chip wiring.

Barcol Hardness A measurement of the surface hardness of non metallic materials in determining the percent of cure. Readings of 40-55 indicate sufficient cure.

Bare Board A printed wiring board which has been sheared to size, printed, etched, plated, and drilled but no components mounted on its surfaces.

Barium Titanate A chemical compound $BaTiO_3$, used in making high dielectric constant ceramic capacitors. Also used in high K thick film pastes.

Barrel That part of a terminal or contact that is crimped.

Barrel Plating A method in which a large number of small parts are placed in a rotating drum and plated.

Barrier A dielectric material which is used to insulate electrical circuits from each other or from ground.

Barrier Layer A thin metal coating which prevents or decreases the growth of intermetallics. For example, a barrier layer of nickel between copper and tin will decrease the growth of tin-copper intermetallics.

Barrier Strip A continuous line of dielectric material used to insulate circuits from each other and from ground.

Base (1) The bottom of an electronic package, usually made of a high thermally conductive material. (2) An insulating support for printed patterns. (3) An alkaline chemical.

Base Material In PWB's, the insulating material that supports the conductive patterns and electronic devices. It can be rigid or flexible.

Base Metal The metal from which metal parts are made that can be electroplated.

Basic Device The simplest useful device which exhibits solid state phenomenon.

Basket Weave In this type of weave, two or more warp threads cross alternately with two or more filling threads. The basket weave is less stable than the plain weave but produces a flatter and stronger fabric. It is also a more pliable fabric than the plain weave and a certain degree of porosity is maintained without too much sleaziness, but not as much as the plain weave.

Batch Processing A manufacturing method in which large numbers of components are processed simultaneously.

Bayonet Coupling A quick coupling device for plug and receptacle connectors.

Beam Lead A long, usually flat, metallic lead which extends beyond the body of a device. One end is attached to the chip device and the other is connected to circuitry on the substrate.

Beam Lead Device An active or passive chip device having beam leads.

Beam Tape A flat, flexible material, made of thin polyimide film and copper foil which is formed into beam leads for TAB bonding.

Belled Mouth A flared or widened entrance to a connector barrel which provides easier insertion of the conductor or mating contacts.

Bellows Contact A connector contact which has a flat spring that is folded which provides a uniform spring loading to the mating part.

Beryllia Beryllium oxide (BeO). A material with high thermal conductivity and used for thick film substrates in high power applications.

Beryllium Copper A metal with superior fatigue properties at high temperatures. Used in applications requiring repeated insertion and extraction.

Betascope A beta particle backscatter gauge which uses radio active isotopes to measure plating thickness.

Bias Voltage The base voltage which establishes a semiconductor's desired d-c operating voltage.

Bifet A linear circuit which combines bipolar and FET transistors together, on the same chip, for improved performance and cost savings.

Bifurcated A lengthwise slotting of a flat spring to provide additional points of contact.

Bifurcate Connector A hermaphrodite connector containing fork-like mating contacts.

Bifurcated Contact A type of contact used in connectors which are made from a flat spring type material and slotted lengthwise thereby providing more points of contact.

Binary A numbering system using only two digits; 1 and 0 and is used for digital IC's.

Binders Materials which are added to compositions to provide strength for handling purposes. Such as binders added to thick film pastes and unfired substrates prior to prefiring.

Binding Post A fixed support to which wire conductors are connected.

Bipolar Device A current driven electronic device with two poles.

Bipolar Transistor A transistor that uses both negative and positive charge carriers.

Birdcage A defect in a stranded wire. A separation from the normal lay of the strands.

Bit Abbreviation for binary digit. The smallest part of information in a binary system. A bit is either a 1 or a 0.

Blackbody A radiator and absorber of radiant energy.

Black Box A loosely used term which is used to describe an electronic assembly chassis within a larger system.

Black Oxide An oxidizing process applied to mating surfaces of a metal package and used for improved sealing properties.

Blade Contact A flat male contact designed to mate with a flat female contact or a tuning fork.

Bleed To give up color when in contact with a liquid medium. In plastic materials, the surfacing of undesired materials.

Bleeding (1) A condition in which a plated hole discharges process material or solution from crevices or voids. (2) The spreading of a thick film paste into adjacent unwanted areas during drying or firing.

Blending The mixing of different viscosities of the same types of thick film pastes, conductive or resistive, to achieve intermediate viscosities or resistivities.

Blind Via A via which extends from one or more inner layers to the surface of a substrate or board.

Blister A raised area on the surface of a molded part caused by the pressure of gases inside on its incompletely hardened surface.

Blistering Localized swelling and separation between any of the layers of the base laminate or between the laminate and the metal cladding. It is a form of localized delamination.

Block (1) A connector housing. (2) To plug up open mesh in a screen to prevent deposition in unwanted areas.

Block Copolymer A compound which results from a chemical reaction between a number of molecules, which are a block of one monomer and a number of molecules which are a block to a different monomer.

Block Diagram A connection diagram in which the essential units are shown in form of blocks and the relationship between blocks are indicated by connecting lines.

Blowhole A hole or void in a solder connection caused by outgassing during soldering.

Blowing Agent Chemicals that can be added to plastics which generate inert gases upon heating. This blowing or expansion causes the plastic to expand, thus forming a foam. Also known as foaming agent.

Blow Molding A method of fabrication of thermoplastic materials in which a parison (hollow tube) is forced into the shape of the mold cavity by internal air pressure.

Boat A container used to hold materials which are to be evaporated or fired.

Bobbin Lugs A lug which serves to connect coil wires to external lead wires. They are usually mounted in plastic bobbins.

Boiling The changing of a liquid into a vapor accompanied by the formation of bubbles within the liquid.

Bolt-Type Connector A type of connector in which contact is made between the conductor and the connector by clamping bolts.

Bomb A chamber in which packages are stored for pressurizing or depressurizing.

Bond (1) The union of materials by adhesives. To unite materials by means of an adhesive. (2) An attachment between a die/substrate or substrate and package using an adhesive for mechanical reasons or an interconnection such as a thermocompression or ultrasonic wire bond to perform an electrical function.

Bondability The surface conditions, such as cleanliness of the bonding areas, that are necessary to provide a reliable bond.

Bond Deformation The change in shape of the lead made by the bonding tool in making the bond.

Bond Envelope The established range of bonding parameters over which acceptable bonds can be made. See Bond Schedule.

Bonding The joining together of components for mechanical, electrical, or sealing functions.

Bonding Area A metalized area at the end of a thin metallic strip or on a semiconductor chip to which a wire bond or electrical connection is made.

Bonding, Die The attaching of a semiconductor chip to a bonding area on a substrate with a conductive or dielectric adhesive or a eutectic or solder alloy.

Bonding Layer An adhesive layer used in bonding other discrete layers during lamination.

Bonding Pad See Bonding Area. Also called Bond Site, Bonding Island, Bond Surface.

Bonding Wire Fine gold or aluminum wire for making electrical connections in monolithic or hybrid circuits between various bonding pads on the semiconductor device substrate and device terminals or substrate lands.

Bond Interface The interface between the gold wire and the bonding area on the substrate.

Bond Lift-Off The failure mode whereby the bonded lead separates from the surface to which it was bonded. Also called bond off.

Bond Parameters See Bond Schedule.

Bond Schedule The pre-established values of the bonding machine parameters used when adjusting for bonding. (Pressure, time, temperature.)

Bond Separation The distance between the attachment points of the initial and second bonds of a wire bond.

Bond Site The part of the bonding area which the wire bond was made.

Bond Strength (1) The unit load applied in tension, compression, flexure, peel, impact, cleavage, or shear, required to break an adhesive assembly with failure occurring in or near the plane of the bond. (2) In wire bonding, it is the strength at the break point of the bond interface measured in grams. (3) The force per unit area required to separate two adjacent layers by a force perpendicular to the board surface; usually refers to the interface between copper and base material.

Bond-to-Bond Distance The distance measured from the bonding site on the die to the bond impression on the post, substrate land, or fingers, which must be bridged by a bonding wire or ribbon.

Bond-to-Chip Distance In beam lead bonding the distance from the heel of the bond to the component.

Bond Tool The device used to position the leads over the designated bonding area and to apply sufficient energy to the leads to form a reliable bond.

Boot A mold placed around the wire terminations of a multiple-contact connector to contain the liquid potting resin during cure.

Boron Fibers High-modulus fibers of elemental boron deposited onto a thin tungsten wire. Supplied as single strands or tapes.

Borosilicate Glass A type of glass having a closely matched coefficient of expansion between metal leads (Kovar) and ceramic packages.

Boss A projection on a plastic part designed to add strength, to facilitate alignment during assembly, to provide for fastenings, etc.

Bow Deviation of a base material or circuit board from flatness.

Box Pattern A pin arrangement for plug-in packages in which the pins are arranged in rows, forming a square or rectangle.

Box Style Wire Contact A design feature in which the wire is completely enclosed in a contact and cannot be pushed through the connector.

Braid A woven metallic wire used for shielding or as a ground wire. Also a woven fibrous protective covering over multiple conductors as in a cable.

Branch Connector A connector which joins a branch conductor to the main conductor at a specific angle.

Brass An alloy of copper and zinc with optional small amounts of tin and lead. Its low cost and excellent electrical conductivity makes it a good candidate for electrical applications.

Braze To join metals with a nonferrous alloy at temperatures above 800 degree F. Also known as hard soldering.

Brazed Terminal A terminal with a barrel seam brazed to form one piece.

Brazing The joining of two similar or dissimilar metals by melting a brazing metal having a lower melting point than the base metals.

Breadboard A prototype version of a circuit to prove the feasibility of a circuit.

Breakaway In screen printing, the distance between the top surface of the substrate and the underside of the screen when the squeegee is not in contact with the screen.

Breakdown The change in properties of a dielectric or insulating material.

Breakdown Voltage The voltage at which an insulator or dielectric ruptures or at which ionization and conduction takes place in a gas or vapor.

Breakout The location in a multiconductor cable at which one or more conductors are separated from the cable to complete circuits to other points.

Bridging The formation of a conductive path between conductors.

Bridging, Solder The filling or bridging of the space between parallel conductors, which are close together, with solder. This could occur on the surface of PWB, downstream from where leads protrude from leaded packages, etc.

B Stage An intermediate stage in the curing of a thermosetting resin. B stage resin can be heated and caused to flow, thereby allowing final curing in the desired shape. Also called Prepreg, for fibers impregnated with B stage resin.

B-TAB See Bumped Tab.

Buffing Stripper A tool that removes flat cable insulation from conductors. Also called Abrasion Stripper.

Bulk Conductance Conductance between two points in a homogeneous material.

Bulk Density The density of a molding material in loose form (granular, nodular, etc.) expressed as a
ratio of weight to volume (e.g., g/cm^3 or lb/ft^3).

Bulk Factor Ratio of the volume of loose molding powder to the volume of the same weight of resin after molding.

Bulkhead Connector A type of connector designed to be inserted in a panel cutout from the component side.

Bulk Resistance That portion of the contact resistance attributed to the length, cross section, and type of material.

Bulk Rope Molding Compound Molding compound made with thickened polyester resin and fibers less than 1/2 inch long. Supplied as rope, it molds with excellent flow and surface appearance.

Bump A small mound or hump that is formed on the chip or the substrate bonding pad and is used as a contact in facedown bonding.

Bumped Chip A chip that has on its termination pads a bump of solder or gold used for bonding to external contacts.

Bumped TAB An acronym for Tape Automated Bonding when the raised solder bump is attached to the tape material.

Bumped Tape A tape which contains the inner lead bond sites as raised metal bumps for the Tape Automated Bonding process.

Buried Resistors Thick film resistors which are located on inner layers of multilayer substrates in order to reduce conductor lengths.

Buried Via A via which connects inner layers but does not extend to the surface of a substrate or board. Sometimes referred to as a Hidden Via.

Burn-In Subjecting electronic components to elevated temperatures for extended periods of time to stabilize their characteristics or cause failure to marginal devices.

Burn-In, Dynamic In addition to burn in; the devices are subjected to actual operating conditions.

Burn-In, Static In addition to burn in; the devices are subjected to a constant voltage rather than the actual operating conditions, either forward or reverse bias.

Burning Rate The tendency of plastic materials to burn at given temperatures.

Burn Off Also called Flame Off. The operation in which the wire is cut by passing a flame across the wire thereby melting it. Used in gold wire thermocompression bonding to form a ball in the subsequent forming of a ball bond.

Bus A wire used to transmit power or data. It can be bare, tinned, or insulated.

Bus Bar A heavy copper or aluminum strip or bar used to carry large amounts of current.

Bus Reactor A current limiting device that is connected between two buses or two sections of one bus to limit any disturbances caused by either bus or bus section.

Bussing The connecting or joining of two or more busses.

Butt The joining of two conductors, end to end, with no overlap and with their axis in line.

Butt Connector A type of connector in which two conductors come together, end to end but do not overlap and with their axis in line.

Butt Contact A configuration in which the mating surfaces engage end-to-end without overlap and their axis in line. The engagement is by pressure with the ends designed to provide optimum surface contact.

Butter Coat A significant thickness of nonreinforced surface-layer resin of the same composition as that within the base material.

Butting Die A crimping die in which the nest and indentor touch at the ends of the crimping cycle.

Butt Joint A connection formed by the end of a lead with the printed wiring board land pattern. (Also called an I-Lead.)

Button-Hook Contact A contact having a curved, hook shaped configuration.

Byte A group of eight bits used to encode a single letter, number, or symbol.

Cable (1) A stranded conductor with or without insulation or cover. (2) Many conductors insulated from each other as a multi-conductor cable.

Cable Assembly A cable equipped with connectors or plugs at each end.

Cable Clamp A clamping device attached at the rear of a receptacle, and around the wires, to provide mechanical support.

Cable Terminal A device which seals the end of a cable and provides insulated egress for the conductors.

Camber The amount of warpage in a substrate.

Cantilevered Contact A spring contact in which the contact force is provided by cantilevered springs. It provides a uniform contact pressure and used in PWB connectors.

Capacitance (capacity) The property of a system of conductors and dielectrics which permits the storage of electricity when potential difference exists between the conductors. Its value is expressed as the ratio of quantity of electricity to a potential difference. A capacitance value is always positive.

Capacitance Density The amount of capacitance available per unit area (pF/mil^2 or mfd/in^2).

Capacitor A device (condenser) whose function is to introduce capacitance into the circuit.

Capillary A hollow tube, used as the bonding tool, and contains the bonding wire. It applies the pressure to the wire during the bonding cycle. (Also known as Capillary Tool.)

Captive Device A multi-part fastener, usually screw type, which retains the loosened components, without separation, when removed from the assembly.

Carbon Tracking A phenomenon wherein a high voltage causes a breakdown on the surface of a dielectric or insulating material and forms a carbonized path.

Carbon-Carbon-Composite Carbon or graphite fibers made into a structural form by weaving, braiding, or other textile techniques and are made dense by adding a carbonaceous matrix. For ultra high temperature applications.

Card Same as Daughter Card.

Card Cages A container, equipped with guide rails, which provides compact packaging of PWB's. Equipped with heat sink devices, connectors, etc. (Also known as Card Racks.)

Card Edge Connector A type of connector that mates with printed wiring conductors which extend to the edge of a printed wiring board. (Also known as a Edgeboard Connector.)

Card Guide A metal or non metal guide which provides easier insertion and extraction of a printed wiring board into or from a connector.

Card On Board A packaging design in which several printed circuit cards are electrically and mechanically connected to printed circuit boards at 90 degree angles.

Card Rack See Card Cages.

Card Slot The lengthwise opening in a printed circuit edge connector that receives the printed circuit board.

Cast To embed a component or assembly in a liquid resin, using molds or shells. Curing or polymerization takes place without external pressure. (See Embed and Pot.)

Castellation A semicircular or crown-shaped metalized surface for making attachment to land patterns as used on ceramics leadless chip carriers.

Catalyst A chemical that causes or speeds up the cure of a resin, but that does not become a chemical part of the final product. Catalysts are normally added in small quantities. (Also know as Hardener, Promoter, Accelerator and Inhibitor.)

Cathode The negative (-) pole in an electroplating bath. Positive (+) charged ions in the plating bath leave the plating bath and deposit on the metal part being plated.

Cavity The depression in a mold, which usually forms the outer surface of the molded part. Depending on the number of such depressions, molds are designated as single-cavity or multicavity molds.

Cell A single unit capable of serving as a d-c voltage source by means of transferring ions during the course of a chemical reaction.

Cellulose A carbohydrate found in plants and used in thermoplastic materials.

Cellulosic Resins A family of resins; cellulose acetate, cellulose acetate butyrate, and cellulose acetate propionate. These are thermoplastic compounds which have good electrical properties.

Centerline Average (CLA) The arithmetical average of measured deviations in a surface profile from a mean centerline located between the peaks and valleys.

Center-To-Center Distance See Pitch.

Center-To-Center Spacing See Pitch.

Centerwire Break In a wire pull test, when the wire fractures at approximately the midpoint between the bonded ends.

Certification A verification that specific testing has been performed and the requirements have been met.

Central Processor (CP) That part of the computer system which houses the main storage, arithmetic unit, and special groups. It preforms arithmetic operations, controls processing, and timing signals. Also called Central Processing Unit.

Centrifugal Casting A fabrication process in which the catalyzed resin is introduced into a rapidly rotating mold where it forms a layer on the mold surfaces and hardens.

Centrifuge (See Constant Acceleration.)

Centipoise
A unit of viscosity, conveniently and approximately defined as the viscosity of water at room temperature. The following table of approximate viscosities at room temperature may be useful for rough comparisons:

Liquid	Viscosity in Centipoises
Water	1
Kerosene	10
Motor oil SAE10	100
Castor oil, glycerin	1,000
Corn syrup	10,000
Molasses	100,000

Ceramic
Inorganic, nonmetallic compounds such as alumina, beryllia and steatite which are fabricated into parts through heat processing.

Ceramic Based Microcircuit
Thick film screened circuits on a ceramic substrate. Circuits usually contain screened resistors, capacitors, and conductors.

Ceramic Chip Carriers
A chip carrier made of alumina, beryllia, or steatite.

Ceramic Matrix Composites
The family of materials similar to carbon-carbon, except the reinforcements are silica-based, such as silicon carbide fibers with a ceramic matrix (silicon carbide, hafnium carbide).

Ceramic Quad Flat Pack (CQFP)
Same as Cerpac.

Cerdip
A ceramic dual-in-line package. It is a package consisting of a metal lead frame which is located between two ceramic layers and subsequently sealed by firing with a glass frit.

Cermet
A solid homogeneous material made of finely ground metal and dielectric particles in intimate contact.

Cermet Thick Film (CTF)
A thick film deposition formed by firing cermet inks or pastes at high temperature onto a ceramic or other high temperature substrates.

Cerpack
A ceramic package with lead frames extending from all four sides and used extensively in surface-mount-technology. Also known as Cerquad, Cerpac, or Cerpak.

Channels
Paths for providing input to and output from computers or other electronic devices.

Charge
(1) In electrostatics, the amount of electricity present on any substance which has accumulated electrical energy. (2) The electrical energy stored in a capacitor or battery.

Chase A structural frame which supports a screen or stencil.

Chassis A metal box or enclosure in which electronic components and printed wiring board assemblies are mounted.

Chemically Deposited Printed Wiring
Printed wiring which is formed on a dielectric base material by the reactions of chemicals.

Chemically Reduced Printed Circuit
A printed circuit which is formed by etching conductor patterns on a metal clad/dielectric base material.

Chemical Reversion The tendency of a cured organic material to soften and return to some other stage other than the cured condition. Caused by humidity, temperature, and pressure or combination thereof.

Chemical Stability The ability of a material to resist change in its structure over an extended period of time by a chemical substance.

Chemical Vapor Deposition A process whereby circuit elements are deposited on a substrate by chemical reduction of a vapor on contact with the substrate.

Chemorheology The flow and chemical reaction of materials in a polymer.

Chessman A term given to a disk, knob, or lever which is used to control the bonding tool with respect to the substrate.

Chip A bare semiconductor die.

Chip And Wire A technology in which semiconductor chips are bonded to substrates and interconnected by wire bonds.

Chip Architecture The design of an IC chip. Contains arithmetic logic units, registers, and other configurations.

Chip Capacitors A chip, made of ceramic or tantalum materials which adds capacitance to a circuit.

Chip Carrier An enclosure or package which is used to house a semiconductor device which has metalized terminations. The metalized terminations on the IC device is subsequently connected to terminations on the chip carrier. See Leaded Chip Carrier and Leadless Chip Carrier.

Chip Circuits Integrated circuits which are interconnected on a substrate to form higher level functions. Usually unpackaged and subsequently sealed in a hermetic package.

Chip Component An IC, diode, transistor, resistor, or capacitor in the form of a chip and used in microelectronic circuits.

Chip Design, Depopulated A gate array or cell array chip design which readily lends itself to automatic wire bonding.

Chip On Board Technology A term describing the mounting of dies directly on substrates and subsequent wire bonding, tape-automated bonding, or flip chip bonding for making electrical interconnections.

Chip-Outs A cavity shaped defect in the surface of a chip leaving an exposed active junction.

Chip Packaging The process of physically locating, connecting and protecting semi-conductor devices in an enclosure.

Chip Resistor A small geometrically shaped ceramic component, approximately 25 - 100 mils, square or rectangular in size. They are made of an inert material with a Ruthenium oxide surface.

Chisel A wedge shaped bonding tool used in wedge and ultrasonic bonding of gold and aluminum wires to elements or package leads.

Chlorinated Hydrocarbon An organic compound having hydrogen atoms and, more importantly, chlorine atoms in its chemical structure. Trichloroethylene, methyl chloroform, and methylene chloride are chlorinated hydrocarbons.

Chopped Bond Wire bonds with gross deformation which greatly reduces the strength of the bonds.

Chopped Fibers Short fibers, usually fractions of an inch in length, chopped from long, continuous fibers. Commonly used as reinforcement in molded plastics.

Chopped Mat Randomly oriented unwoven fibers cut to various lengths and compacted together by heat and pressure. Fibers are normally several inches long. Commonly used in making glass-reinforced sheets or sheet forms.

Chopped Roving Chopped sections of roving. (See Roving.)

Chuck That part of the bonding equipment which holds the unit to be bonded.

Circuit The interconnection of electrical elements and devices to perform a desired electrical function.

Circuit Board Packaging The design and use of chips on printed circuit boards.

Circuit Element Any basic component of a circuit, excluding interconnections.

Circuit Layout The positioning of the conductors and components prior to photoreduction of the layout to obtain a positive or negative.

Circuit Verifier A test analyzer which electrically stimulates a device and monitors it for shorts, opens, etc.

Circular Mil A unit of area equal to the area of a circle whose diameter is 1 mil. (0.001 inches). Used in specifying cross sectional area of round wires.

Circumferential Crimp A type of crimp in which a force is exerted around the entire circumference of a terminal barrel by the crimping dies and forms symmetrical indentations.

Circumferential Separation (1) A crack or separation in the plating around the circumference of a plated-through-hole. (2) In the solder fillet around the wire lead or an eyelet. (3) Between a solder fillet and a land.

Clad A condition of the base material, to which a relatively thin layer or sheet of metal foil (cladding) has been bonded on one or both of its sides. The result is called a metal-clad base material.

Cladding (1) In fiber optics, a sheathing of a lower refractive index material around the core of a higher refractive index material which provides optical insulation and protection to the reflective interface. The surface layer of sheet or foil in clad materials.

Clamping Force The force applied to a bonding tool to effect a bond.

Clamping Screw A threaded screw or bolt located in the terminal block which, when tightened, compresses the wire or conductor against the current bar.

Clean Room A dedicated manufacturing area where the air is filtered and the dust particle size and quantities are controlled to various levels and the temperature and humidity also controlled.

Clearance The shortest distance between two lines, objects, tracks, etc.

Clearance Hole A hole in the conductive pattern larger than, but coaxial with, a hole in the printed circuit base material.

Clinched Lead A wire which is bent so that a spring action is formed against the mounting holes to make electrical contact with the conductive pattern prior to soldering.

Clinched-Wire-Through Connection An electrical connection made by a component lead or wire after it is inserted through a hole in a printed circuit board and formed or shaped to provide contact with the conductive patterns on both sides of the board and subsequently soldered in place.

Clip Terminal The point where the hook-up wire is clipped against the connector post.

Clock A term used in electronic systems which assures that IC's are turned on and off simultaneously.

Clocking The arrangement of connector inserts, jack screws, pins, sockets, keys, keyways, etc. to eliminate mismating or cross mating of connectors.

Closed End Splice When two or more conductors enter the same end of a barrel terminal.

Closed Entry A design that limits the size of mating parts.

Closed Entry Contact A female contact designed to prevent the entry of a pin having a larger diameter than the mating pin.

Coat To cover with a finishing, protecting, or enclosing layer of any compound.

Coated Metal Core Substrate A composite substrate which has a metal base which is coated with a dielectric. The outer surfaces are subsequently metal coated and circuits formed.

Coating A thin layer of material, conductive or dielectric, that is applied over components or the base material.

Coaxial Cable A cable consisting of a center conductor surrounded by a dielectric and an outer conductor or shield which protects external radiation from affecting the current flowing in the center conductor.

Co-Axial Controlled Impedance When the impedance is essentially constant along the entire length of two or more insulated conductors.

Coefficient Of Thermal Expansion (CTE) The ratio of the change in dimensions of a material to the change in temperature. The CTE of a material is usually linear but can be volumetric. It is expressed as a factor of 10^{-7} cm/cm/degree Celsius.

C-1010 (CRS)	140
Kovar	55
Alloy 52	90
Glass	55
Alumina	50
Beryllia	70
Copper	160
Stainless Steel	160

Co-Fire A process in which unfired ceramic and screened circuits are fired simultaneously.

Cohesion The state in which the particles of a single substance are held together by primary or secondary valence forces. As used in the adhesive field, the state in which the particles of the adhesive (or the adherend) are held together.

Coil Form Terminals Terminals which are mounted on the coil base to which the coil wires are connected.

Coined In thick film, a screen which contains the impression of a substrate. It is caused by abuse due to incorrect printer set up parameters.

Cold Flow The continuing dimensional change that follows initial instantaneous deformation in a nonrigid material under static load. Also called Creep.

Cold Pressing A bonding operation in which an assembly is subjected to pressure without the application of heat.

Cold-Press Molding Molding process where inexpensive plastic male and female molds are used with room-temperature-curing resins to produce accurate parts. Limited runs are possible.

Cold Short Brittleness in metals at temperatures below recrystallization.

Cold Solder Connection A soldered joint that was made with insufficient heat or the parts being soldered moved during solidification. It exhibits a poor wetting and grayish, porous appearance.

Cold Weld A joining of two metals achieved by pressure alone.

Cold Work The embrittlement of a metal caused by repeated flexing actions.

Collector The element in a transistor which collects the current generated at the junction between the emitter and base. The output part of the transistor.

Collector Electrode A metalized bonding pad in contact with the collector of a transistor element.

Collector Junction The junction between the collector and base in a transistor.

Collimated Light Parallel rays of light.

Colophony A natural product (rosin) whose constituents originates in nature as raw materials. See Flux and Rosin.

Colorant A dye which is added to a resin for coloring purposes.

Color Coding The marking of wires, terminals, or contacts with a color to aid in identification.

Comb Pattern A test pattern applied to a substrate in the form of a comb.

Common Part A part which may be used on two or more major items.

Compatible Materials which can be mixed together or brought into contact with each other with little or no reaction or separation from one another.

Compensation Circuit A circuit which alters the functioning of another circuit to which applied with the goal of achieving a desired performance; temperature and frequency compensation are the most common.

Complementary Metal Oxide Semiconductor (CMOS)
A device composed of p and n type MOS used to achieve low power consumption.

Complementary Transistors Two transistors of opposite conductivity, (pnp and npn) in the same functional unit.

Complete Solder Joint A solder joint free of any defects such as nonwetting areas, voids, and improper fillets.

Compliant Bond A bond which uses an elastically and/or plastically deformable member to impart the required energy to the lead. This member is usually a thin metal foil that is expendable in the process.

Compliant Member The elastically and/or plastically deformable medium which is used to impart the required energy to the lead(s) when forming a compliant bond.

Component Has many meanings such as an active or passive device or element, integrated or functional circuit, or part of a system.

Component Density The number of components on a substrate, PWB, etc. per unit of area.

Component Hole A hole in a printed circuit board used for inserting component leads, pins, and wires.

Component Side The side of a PWB on which the components are to be mounted or are mounted. Also called the Primary Side.

Composite (1) A homogeneous material created by the synthetic assembly of two or more materials (a selected filler or reinforcing elements and compatible matrix binder) to obtain specific characteristics and properties. (2) Combinations of materials, as opposed to single or homogeneous materials. See Organic Composites, Carbon-Carbon Composites, Metal Matrix Composites and Ceramic Matrix Composites.

Compound Some combination of elements in a stable molecular arrangement.

Compression Connector A connector which, after the ends of the conductors have been inserted in a tube like body connector, is crimped by an external force, resulting in the crimping of the connector and the ends of the conductors.

Compression Molding A technique of thermoset molding in which the molding compound (generally preheated) is placed in the heated open mold cavity and the mold is closed under pressure (usually in a hydraulic press), causing the material to flow and completely fill the cavity, with pressure being held until the material has cured.

Compression Seal A seal between an electronic package and its leads. The seal is formed as the heated metal, on cooling, shrinks around the glass seal forming a tight joint.

Compressive Strength The maximum compressive stress a material is capable of sustaining. For materials which do not fail by a shattering fracture, the value is arbitrary, depending on the distortion allowed.

Computers Electronic machines which can accept, store, and process information mathematically according to instructions, and subsequently provide results after processing the information.

Condensation A chemical reaction in which two or more molecules combine with the separation of water, or some other simple substance. If a polymer is formed, the process is called polycondensation.

Condensation Resins Any of the alkyd, phenol-aldehyde, and urea-formaldehyde resins.

Conductance The reciprocal of resistance. It is the ratio of current passing through a material to the potential difference at its ends.

Conduction, Thermal The flow of heat in a material from a hot region to a cooler region.

Conductive Adhesive, Electrical An adhesive to which metal particles are added to increase the electrical conductivity. Usually silver particles but sometimes gold or copper.

Conductive Contaminant Growth The electrical shorting of circuits by conductive salts from plating baths, etching solutions, and solder flux residue due to inadequate water rinsing or flux removal.

Conductive Epoxy An epoxy resin to which has been added metallic particles or thermally conductive (dielectric) powders to increase the electrical or thermal conductivity.

Conductive Pattern A design formed from an electrically conductive material on an insulating base.

Conductivity The ability of a material to conduct electrical or thermal energy.

Conductor, Electrical A material that is suitable for carrying electrical current or having low resistivity ($<10^{-4}$ ohms per cm).

Conductor Spacing The distance between adjacent edges of printed circuit lines or patterns.

Conductor Stop A protective device or tool which limits the extension of a wire beyond the conductor barrel.

Conductor To Hole Spacing The distance between a conductor edge and the edge of a conductor hole.

Conductor Width The width of individual conductors or lines in a conductive film pattern.

Configuration The relative arrangement of parts; such as the components in a circuit.

Confined Crescent Crimp Two crescent shaped configurations; one is located at the top and the other at the bottom of the wire barrel crimp.

Conformal Coating A thin dielectric coating applied to the circuitry and components of printed wiring assemblies for environmental protection against moisture as well as fingerprints. Also provides mechanical protection.

Connection A point in a circuit where two or more components are joined together and has nearly zero impedance.

Connection Diagram A diagram showing the electrical connections between the parts and circuits.

Connections (1) All the electrical attachments of the nets interconnecting logic units on a package level that are connected to the next higher package level. (2) A part of a circuit which has zero impedance that joins components, devices, etc., together.

Connector A part which provides ease in electrical connections and disconnections.

Connector Area The area of a printed wiring board used for providing external electrical connections.

Connector Assembly A mated plug and receptacle.

Connector Block Same as Connector Housing.

Connector Housing A molded block of insulating material containing contacts. Sometimes called Connector Block.

28

Connector Insert An electrically insulated molded part which holds the contacts in proper arrangement and alignment.

Connector Insertion Loss The amount of power lost due to the insertion of a mated connector onto a cable. Also called Coupling Loss.

Connector Module A group of connector inserts which have the same outside dimensions and can accept different types of contacts and have different contact shapes.

Connector Patch The distance from the center of one conductive layer to the center of the adjacent conductive layer.

Connector Set The plug and receptacle connectors designed to be used or mated together.

Connector Shell The casing which holds the connector insert and contact assembly. They provide proper alignment and protection of projecting contacts.

Constant A permanent, fixed, or unvarying value. Usually designated by K.

Constant Acceleration A means of physical testing the integrity of wire bonds, soldered connections, and adhesive bonds in hybrid packages by centrifuging (spinning) at high rates of speeds (5,000 - 10,000 RPMs) which imparts a **g** loading on bonds and bonded elements. Also called Centrifuge.

Constraining Core Substrate A composite printed wiring board composed of epoxy-glass outer layers which are bonded to a low thermal-expansion core material such as graphite-epoxy, aramid fiber-epoxy, or copper-invar-copper. The low thermal expansion core material constricts the expansion of the outer layer.

Constriction Resistance That part of the contact resistance which is due to the contact to circuit board interface.

Contact The current carrying member of a connector that engage or disengage to open or close circuits.

Contact Alignment The amount of excess space which contacts have within the cavity so as to permit self-alignment of mated contacts.

Contact Angle In a solder joint, it is the angle of wetting between the fillet and the land or pad. Contact angles of less than 90 degrees are acceptable. Those which are greater than 90 degrees are unacceptable.

Contact Area The mating surface area between two conductors or a conductor and a connector through which current flows.

Contact Arrangement The number, spacing, and pattern of contacts in a connector.

Contact Back Wipe During the actuation cycle, the contacts travel along the mating surfaces and then return on a cleaned wiped surface at the end of the actuation cycle.

Contact Cavity A defined hole in a connector insert in which the contacts must fit.

Contact Chatter The vibration of mating contacts resulting in the opening and closing of the contacts.

Contact Durability A measure of the number of insertions and withdrawal cycles that a connector withstands while meeting its specifications.

Contact Engaging And Separating Force The force required to engage and/or separate pins and sockets type contacts when they are not physically located in the connector inserts. Usually minimum and maximum values are given.

Contact, Female A contact molded in the body such that the mating portion is inserted into the unit. Similar in function to a Socket Contact.

Contact Inspection Hole A hole located in the rear part of a contact used to determine the depth to which a wire has been inserted.

Contact Length The length of travel of one contact while touching another contact during assembly or disassembly.

Contact Male A contact located in the body such that the mating portion extends into the female contact. Similar to a pin contact.

Contact Plating The metal plated on the base material of a contact to provide wear resistance.

Contact Positions The total number of contacts in a connector. However, in an edge type connector, it is the number of contact positions along the length of the connector.

Contact Pressure The force which mating surfaces exert on each other.

Contact Printing In screen printing when the screen is in contact with the substrate.

Contact Resistance (1) In leaded devices, the apparent resistance between the terminating lead and the body of the device. (2) The maximum resistance allowed between a pin and socket contacts of a connector when assembled and in use.

Contact Retainer A device which holds or houses the contact.

Contact Retention The minimum amount of force required of a contact to remain engaged within the connector insert.

Contact Shoulder The raised part of the contact which limits its travel distance into the insert.

Contact Size A designation which specifies the wire gage size and the diameter of the engagement end of the pin.

Contact Spacing The centerline distance between adjacent contact areas.

Contact Spring A spring which is placed inside the socket-type contact during its manufacture which exerts a force on the pin to insure intimate contact.

Contact Wipe The contact area over which mating contacts surfaces touch during insertion and separation.

Contaminant (1) An impurity in a material which can affect its properties. (2) Undesirable particulate which can adversely affect the quality of the product.

Continuous Current Rating The stated rms alternating or direct current which a connector or any other electrical device can carry on a continuous basis under specified conditions.

Continuous Belt Furnace A furnace equipped with a continuous belt for transporting unfired substrates through a firing cycle.

Continuous Use The operation of a component, device, or system for an indefinite period of time.

Continuity A continuous path for the flow of current in an electrical circuit.

Control Cable A multilead flexible cable intended for carrying signal circuits only where the current requirements are minimum.

Control Chart A means of recording the performance of a process over a period of time and used to identify problems.

Controlled Collapse Chip Connection A solder joint between a substrate and a flip chip whose height is controlled by the surface tension of the liquid solder.

Controlled Impedance Cable A cable containing two or more insulated conductors in which the impedance is nearly constant along its entire length.

Controlled Part A component or device which is made under controlled manufacturing processes and purchased to specified requirements.

Controlling Collapse Controlling the height of the solder joint after the solder balls have melted when soldering flip chip devices to a substrate.

Convection A transfer of heat or electricity by moving particles of matter.

Convention A prescribed method used in making electronic diagrams so as to illustrate a clear representation of the circuit function.

Cooling The lowering of the temperature of an object or material.

Coordinatograph A precision drafting machine used to make original artwork for integrated circuits.

Coplanarity The distance between the lowest and highest pin in a package when the package is placed on a flat surface.

Coplanar Leads Flat or ribbon type leads extending from the sides of an electronic package all lying in the same plane.

Copolymer A compound resulting from the chemical reaction of two chemically different substances. The resulting compound has different properties than either of the initial substances.

Cordierite A vitreous ceramic material composed of $(2MgO-2Al_2O_3-5SiO_2)$.

Cordwood Module A high density package formed by stacking electronic components between two sheets of film or other dielectric materials and interconnecting them into electrical circuits by welding or soldering the leads.

Core In fiber optics, it is the light conducting, center portion of the fibers bounded by cladding. It also is the high refractive index region.

Corona An electrical discharge, sometimes luminous, due to ionization of the air, appearing on the surface of a conductor when the potential exceeds a certain value.

Corona Resistance The length of time that a dielectric material withstands the action of a specified level of field intensified ionization that does not result in the immediate complete breakdown of the insulation.

Corrosion A chemical action which causes gradual deterioration of the surface of a metal by oxidation or chemical reaction.

Corrosive Fluxes A flux containing inorganic acids and salts which are needed to prepare some surfaces for rapid wetting by the molten solder.

Coulomb The quantity of electricity which passes any point in a circuit in one second when one ampere of current is applied.

Coupled Noise See Cross Talk.

Coupler (1) A component which is used to transfer energy from one circuit to another. (2) A chemical used to improve the bond between a resin and the glass fibers in a composite material.

Coupling Capacitor Any capacitor used to couple two circuits together. It blocks dc signals and allows high frequency signals to pass between parts in an electrical circuit.

Coupling Loss See Connector Insertion Loss.

Coupling Ring A device used on cylindrical connectors to lock the plug and receptacle together.

Coupling Torque The force required to rotate a coupling ring when engaging a mating pair of connectors.

Cover A part of a connector which is designed to cover the mating end for mechanical and environmental protection.

Cover Coat Usually refers to coatings applied to flexible circuits; may be liquid or film.

Cratering A depression or pit under an ultrasonic bond on a chip which was torn loose due to excessive energy transmitted through the wire bond.

Crazing (1) A base material condition in which connected white spots or crosses appear on or below the surface of the base material. They are due to the separation of fibers in the glass cloth and connecting weave intersections. Similar to Measling. (2) Fine cracks which may extend in a network on or under the surface or through a layer of a plastic material.

Creep The dimensional change with time of a material under load, following the initial instantaneous elastic deformation; the time-dependent part of strain resulting from force. Creep at room temperature is sometimes called "Cold Flow". See Recommended Practices for Testing Long-Time Creep and Stress-Relaxation of Plastics under Tension or Compression Loads at Various Temperatures, ASTM D675.

Creepage The conduction of electricity across the surface of a dielectric material.

Creep Distance The shortest distance between conductors measured on the surface of a dielectric material.

Crimp To compress or deform a connector barrel around a wire or cable in order to make an electrical connection.

Crimp Contact A contact whose back portion is a hollow cylinder into which a bare wire can be inserted. Often called a Solderless Contact.

Crimper The part of a crimping die which indents or compresses the terminal barrel.

Crimping Chamber That part of the crimping tool which crimps the contact or terminal.

Crimping Die The part of the crimping tool which shapes the crimp.

Crimping Tool A mechanical device used for crimping contacts and terminals.

Crimp Terminal The location where the crimp was made with the bare wire and the pin or contact which mates with the contact terminal.

Crimp Termination A connection in which a metal sleeve is secured to a conductor by mechanically crimping the sleeve with pliers, or a crimping machine.

Critical Item A part whose failure to meet its designed requirements results in the failure of the product or system. Also called Critical Part.

Cross Connector A connector which joins two branch conductors to the main conductor. The branch conductors are opposite to each other and perpendicular to the main conductor.

Cross Crimp A crimp which applies pressure to the top and bottom of a terminal barrel without collapsing the sides.

Cross-Linking The forming of chemical links between reactive atoms in the molecular chain of a plastic. It is cross-linking in the thermosetting resins that makes the resins infusible.

Crossover A point where one conductive path crosses another. The two paths are insulated from each other by a dielectric layer.

Crosshatching The breaking up of large conductive areas by using a pattern of voids to achieve a shielding effect.

Cross Talk Undesirable electrical interference caused by the coupling of energy between signal paths. Also known as Coupled Noise.

Crow's Foot In this type of weave there is a 3 by 1 interlacing. That is, a filling thread floats over the three warp threads and then under one. This type of fabric looks different on one side than the other. Fabrics with this weave are more pliable than either the plain or basket weave and, consequently, are easier to form around curves.

Crystalline Melting Point The temperature at which the crystalline structure in a material is broken down.

Crystallinity A state of molecular structure referring to uniformity and compactness of the molecular chains forming the polymer and resulting from the formation of solid crystals with a definite geometric pattern. In some resins, such as polyethylene, the degree of crystallinity indicates the degree of stiffness, hardness, environmental stress-check resistance, and heat resistance.

Crystallization The formation of solids having a definite geometric form and the growth of large crystals from smaller ones.

C-Stage The cured or final stage of a resin in which the material will not melt when heated. It may soften, however.

C-Stage Epoxy Glass The cured stage of an epoxy resin impregnated glass cloth composite.

Cull Material remaining in a transfer chamber after the mold has been filled. Unless there is a slight excess in the charge, the operator cannot be sure the cavity is filled. The charge is generally regulated to control the thickness of the cull.

Cumulative Distribution Function (CDF) The division of a parameter as a part of the total number of measurements with respect to a statistic.

Cure To change the physical properties of a material (usually from a liquid to a solid) by chemical reaction, by the action of heat and catalysts, alone or in combination, with or without pressure.

Curie Temperature (Curie Point) Above a critical temperature, ferromagnetic materials lose their permanent spontaneous magnetization and ferroelectric materials lose their spontaneous polarization.

Curing Agent A material which activates a catalyst, already present in a resin material, thereby causing a chemical reaction which results in hardening of the entire mass. Synonymous with Hardener. Also, any chemical which reacts with a base resin, resulting in a final cured or hardened part.

Curing Cycle The total time at a temperature or temperatures which is required to cure or harden a resin or adhesive to achieve maximum properties.

Curing Temperature The temperature at which a material is subjected to curing.

Curing Time In the molding of thermosetting plastics, the time required for the material to be properly cured.

Curls Extruded material extending from the edge of the bond.

Current The rate of transfer of electricity. It is measured in amperes and represents the transfer of one coulomb per second.

Current Carrying Capacity The maximum current that can be continuously carried without causing degradation of electrical or mechanical properties.

Current Mode Logic (CML) Integrated circuit logic in which transistors are paralleled to eliminate current drain.

Current Penetration The depth to which current will penetrate into the surface of a conductor at a given frequency.

Current Rating The maximum continuous amount of current a device/component is designed to carry for a specified time at a specified operating temperature.

Custom Design In electronic packaging, it is a design in which the placement of devices and routing of conductors will vary from a standard array but within specific tolerances.

Cut And Strip A method of making artwork using a two-ply laminated plastic sheet. By cutting and stripping off the unwanted part of the opaque layer from the translucent layer, leaving the desired artwork configuration.

Cycle One complete operation of a molding press from closing time to closing time.

Cycles Per Instruction The number of cycles necessary to process an instruction.

Cycle Time The unit of time that elements of the central processing unit require to complete their functions. This may vary from one or more cycles to complete a function.

Damage The failure of an electronic or mechanical component or a material which require replacement.

Damping In a material, the ability to absorb energy to reduce vibration.

Daughter Card A card which interfaces with a motherboard or backplane (Synonymous with Daughter Board or Daughter Substrate).

DC A circuit or voltage source in which the voltage remains constant irrespective of time.

DC Voltage Coefficient A measure of changes in the primary characteristics of a circuit element as a function of the voltage stress applied.

Dead Face A method used to protect contacts when not engaged.

Dead Front The mating surface of a connector designed so that the contacts are recessed below the surface of the connector insulator body.

Debugging The elimination of early failures by aging or stabilizing the equipment prior to final test.

Decibel (DB) (1) A unit of change in sound or audio intensity. (2) In fiber optics, a unit used as a logarithmic measure to describe the attenuation (optical power loss per unit length) in a fiber.

Decorative Laminates High-pressure laminates consisting of a phenolic-paper core and a melamine-paper top sheet with a decorative pattern.

Decoupling Capacitor A device which filters out transients in a power distribution system.

Defect Any non-conforming, unacceptable characteristic in a unit which requires attention and correction.

Definition The sharpness of pattern edges in printed circuits.

Deflashing Covers the range of finishing techniques used to remove the flash (excess unwanted material) from a plastic part; examples are filing, sanding, milling, tumbling, and wheelabrading. (See Flash.)

Deflection Temperature Formerly called Heat-Distortion Temperature (HDT).

Degradation A gradual deterioration in performance. Sometimes called Drift in electronic equipment.

Deionized Water Water which has been treated to remove ionized material.

Delamination A separation between any of the layers of the base laminate or between the laminate and the metal cladding originating from or extending to the edges of a hole or edge of the board.

Delid The mechanical steps required to remove a lid or cover from a sealed hybrid package.

Demineralized Water Water which has been treated to remove minerals which are normally found in hard water.

Denier The weight of a yarn which determines its physical size or cross section unit or measurement.

Dendritic Growth A conductive tree-like growth which occurs between conductors, usually under the combined influence of electrical energy and humidity. The result is a short between the bridged conductors.

Density The weight of a material in relationship to its volume. Expressed in grams per cubic centimeter or pounds per cubic foot, etc.

Deposited To lay down or to be formed by deposition. Examples are: Vapor Deposition, Sputtering, Electroless, Electrolytic Plating.

Deposition A process of applying a material to a base material by means of vacuum, chemical, screening, or vapor methods.

Depth Of Crimp The distance from the outside diameter of the barrel to the bottom of the indentation made by the indentor. See "T" dimension.

Design (1) To create original sketches, plans, drawings in order to achieve a specific end. (2) The arrangement of parts, details, to produce a complete unit.

Desoldering A process of disassembling soldered parts. Methods used include wicking, solder sucking, and solder extraction.

Detent A bump, pin, or raised projection from the surface of a part.

Detritus Fragments of loose material produced during resistor trimming which remain in the trimmed area.

Device A single discrete electrical element, such as a transistor or resistor, which cannot be further reduced without destroying or eliminating its function.

Devitrification A conversion process of a glassy matter into a crystalline condition.

Dewetting A condition in which liquid solder has not adhered intimately to an area and has pulled back from the base metal leaving the base metal exposed.

Dew Point The temperature at which water vapors in the air begins to condense.

Diallyl Phthalate An ester polymer resulting from reaction of allyl alcohol and phthalic anhydride.

Dice The plural for Die.

Die A tiny uncased integrated device obtained from a semi-conductor wafer. Also called Chip. Examples: transistors, diodes, I.C.'s.

Die Bond A process in which a semi-conductor chip is attached to a substrate by an epoxy or gold-silicon eutectic solder alloy.

Dielectric Any insulating medium or material that does not conduct electricity.

Dielectric Absorption The accumulation of electric charges within the body of an imperfect dielectric material when placed in an electric field.

Dielectric Breakdown The voltage required to cause an electrical failure of the insulation.

Dielectric Constant That property of a dielectric which determines the electrostatic energy stored per unit volume for unit potential gradient. Same as Pevittivity or Specific Inductive Capacity.

Dielectric Layer A layer of dielectric material between two conductor layers.

Dielectric Loss The time rate at which electric energy is transformed into heat in a dielectric when it is subjected to a changing electric field.

Dielectric-Loss Angle The difference between 90 degree and the dielectric-phase angle. Same as Dielectric-Phase Difference.

Dielectric-Loss Factor The product of dielectric constant and the tangent of dielectric-loss angle for a material. (Dielectric-Loss Index).

Dielectric-Phase Angle The angular difference in phase between the sinusoidal alternating potential difference applied to a dielectric and the component of the resulting alternating current having the same period as the potential difference.

Dielectric Power Factor The cosine of the dielectric-phase angle (or sine of the dielectric-loss angle).

Dielectric Strength The voltage which an insulating material can withstand before breakdown occurs, usually expressed as a voltage gradient (such as volts/mil).

Dielectric Withstanding Voltage The maximum electrical stress (in volts) that a dielectric material can withstand without failure.

Differential Scanning Colorimetry (DSC) A technique which measures the physical transitions of a polymeric material as a function of temperature and comparing it to another similar material, undergoing the same temperature cycle but the second material not experiencing any transitions or reactions.

Diffusion (1) A movement of matter at the atomic level from regions of high concentration to regions of low concentrations. (2) A thermal process by which minute amounts of impurities are impregnated and distributed into semiconductor material.

Digital-to-Analog-Converter A device which converts an input digital signal into a proportional analog output voltage or current.

Diluent An ingredient usually added to a formulation to reduce the concentration of the resin or reduce its viscosity.

Dimensional Change Any change in length, width, or thickness of a solid material.

Dimensional Stability A measure of dimensional change caused by temperature, humidity, age, stress and chemical treatment. (Expressed as units/unit.)

Diode A semiconductor device with an anode and cathode which permits current to flow in one direction and inhibits current flow in the other direction.

Diphenyl Oxide Resins Thermosetting resins based on diphenyl oxide and possessing excellent handling properties and heat resistance.

Dip Soldered Terminals All the terminals of a connector which after being inserted in the holes of a printed wiring board are subsequently soldered in position.

Dip Soldering A process in which component leads, that are to be soldered to a PWB, are brought in contact with the surface of molten solder and soldered to the conductive paths on the PWB.

Direct Access Storage Device (DASD) Computer hardware which utilizes magnetic recording on a rotating disk surface.

Direct Capacitance When the capacitance between two conductors is measured through a dielectric layer.

Direct Chip Attach (DCA) The bonding or attachment of a chip to a substrate.

Direct Contact A contact which is made to a semiconductor chip when the wire is bonded directly over the part to be electrically connected rather than a lateral path or by an expanded contact.

Direct Emulsion When a liquid form of emulsion is applied to a screen as opposed to a film type emulsion.

Direct Metal Mask A metal mask which is made by chemically etching a pattern into a thin sheet of metal.

Direct Mounting The attachment of terminal blocks by solder mounting using bottom terminals.

Disconnect A feature of a current carrying device to have the capability of being separated from its mating part.

Discontinuity A separation or interruption resulting in the permanent or temporary loss of current or voltage.

Discrete Component A component which is fabricated prior to installation as opposed to those which are screen printed or vacuum deposited as part of the film network. (Resistors, capacitors, diodes, transistors.)

Dispersion Very fine particles in suspension in another substance.

Displacement Current A current which exists in addition to ordinary conduction current in an a-c circuit. It is proportional to the rate of change of the electric field.

Disruptive Discharge The sudden and large increase in current through an insulation medium due to the complete failure of the medium under the electrostatic stress.

Dissipation The loss of electrical energy generally in the form of heat.

Dissipation Factor The tangent of the loss angle of the insulating material.

Dissolution Rate See Dissolution, Soldering.

Dissolution, Soldering The preferred term is "dissolution rate" which means the rate at which the solid base metal dissolves.

Distance to Neutral Point (DNP) With the neutral point being the geometric center of an array of pads on a substrate and point at which there is essentially motion of the chip and substrate in the x-y plane during thermal excursions, the DNP is the distance from the neutral point to the separation of a joint on a chip. This dimension controls the strain on the joint caused by the mismatch between the chip and the substrate.

Disturbed Solder Joint A joint in which the members to be soldered together were moved during solidification of the solder.

Doctor Blade A straight edge or knife located above the casting slurry to control the thickness of the slurry. The up and down movement of the doctor blade is controlled by a micrometer which in turn controls the thickness of the slurry.

Doping The addition of selective impurities to semiconductor materials to alter its conductivity. Common dopants are aluminum, antimony, arsenic, gallium and indium.

Dosimeter A device which people wear who work in a environment containing radioactive material that indicates levels of radiation to which they are exposed.

Dot Coding A tool imprinting process which indicates whether the proper tool has been used.

Double-Grip Terminal A double crimp, one over the wire and the other over the insulation, to prevent strain from reaching the barrel crimp. Used in vibration applications.

Double-Pole A contact arrangement having two separate contact combinations.

Double Sided Substrate A substrate having circuitry on both the top and bottom sides and which are interconnected by metalized through-holes or edge metalization or a combination of both.

Drag Soldering A process in which the joints of electronic assemblies to be soldered are brought in contact with the surface of a pool of molten solder.

Drain Wire An uninsulated wire in a cable which is in contact with the shield along its entire length and used for terminating the shield and grounding purposes. Also called a Drain Conductor.

Drawbridging A condition which occurs during solder reflow, when chips move and resemble a drawbridge. Same as Tombstone and Manhattan Effect.

Dressed Contact A contact with a locking spring permanently attached.

Drift The rate of change in value of a passive component. Expressed in percentage per 1,000 hours due to the effects of temperature, aging, humidity, etc.

Driver An electronic circuit or separate chip which supplies signal voltage and current or input to another circuit.

Dross Oxides and other impurities which form on the surface of molten solder.

Dry To change the physical state of an adhesive on an adherend through the loss of solvent constituents by evaporation or absorption, or both.

Dry Circuit A circuit in which the voltage and current are at a very low level thereby eliminating any arcing to roughen any contacts.

Dry Film Resist A photo resist material in the form of a light sensitive film. Supplied in roll form and is processed dry.

Dry Inert Atmosphere An atmosphere which has had the water molecules removed by the circulation of an inert gas such as nitrogen.

Dry Pressing The compacting of dry powders with binders in molds under heat and pressure to form a solid mass and subsequently sintered.

Dry Print Screened resistors and conductors which have been processed through the drying cycle, removing the solvents from the thick film paste.

Dual In Line Package (DIP) An electronic device containing an integrated circuit chip which is connected to terminals or leads that are positioned in two straight rows on the sides of the package.

Dual-Wave Soldering A soldering process which utilizes two waves; a turbulent wave and a laminar wave. The first wave, turbulent, which gets into small constricted areas to ensure good coverage while the second wave, laminar, removes solder projections.

Ductility The ability of a material to deform plastically before fracturing.

Dummy Connector Plug A part included in the total design of a connector to mate with a connector but not to perform an electrical function.

Dummy Connector Receptacle A part included in the total design of a connector which mates with a plug connector but cannot be attached to a cable. It is used to seal out moisture.

Dummying A process in which a metal plate having a large total area is placed in an electroplating solution, attached to the cathode, and used for removing impurities from the solution.

Duty Cycle The specified operating and non-operating time of equipment.

Durometer An instrument used to measure the hardness of a non metal such as rubber, plastic, etc.

Dye Penetrant A liquid material containing fluorescent particles used to detect cracks and gross leaks.

Dynamic Flex Flexible circuitry which is used in applications where continuous flexing is required.

Dynamic Gap In a edge type connector, it is the minimum distance between opposite contacts when the printed wiring board is rapidly removed. The dynamic gap is required to prevent an electrical short from occurring.

Dynamic Mechanical Analysis (DMA) A test method for resin type materials in which dimensional changes are measured in the stiffness of prepregs and laminates with changes in temperature.

Dynamic Printing Force The fluid force which causes a thick film paste to flow through a screen mesh and wet the surface below.

Dynamic Random Access Memory (DRAM) The main electronic memory storage system of large computers, minicomputers, and some microcomputer memory systems. It periodically requires refresh cycles to restore and maintain information because it employs transient phenomena, typically charge stored in a leaky capacitor.

Dynamic Testing Testing of hybrid circuits where reactions to a-c, particularly high frequencies, are evaluated.

Eccentricity The center of a conductor's location with respect to the circular cross section of insulation. It is expressed as a percentage of displacement of one circle within the other.

Edgeboard Connector A connector that mates with printed circuits running to the edge of a printed circuit board. (Also called Card Edge Connector or Edge Connector.)

Edge-Board Contacts A series of contacts printed on or near an edge of a printed wiring board and intended for mating with a one-part edge connector.

Edge Card A printed wiring board having a series of contacts printed on or near any edge or side for mating with an edgeboard connector.

Edge Dip Solderability Test A solderability test in which a specimen is fluxed with a non-activated rosin flux and immersing it in a solder pot at a specific immersion rate for a specific time and withdrawing it at a specific rate.

Edge Seal A plug-in package that has a flanged header and cover with a flangeless edge.

Egg Crating The insulated walls between each cavity within the contact wire entry face of the connector housing. Looks like rectangular cells and minimizes danger of shock.

E Glass A low alkali alumina borosilicate glass with excellent dielectric properties.

Eight Harness Satin A fabric with this type of weave has a 7 by 1 interlacing in which a filling thread floats over 7 warp threads and then under 1. Like the crowfoot weave, it looks different on one side than on the other. This weave is more pliable than any of the others and is especially adaptable where it is necessary to form around compound curves, such as on radomes.

Elasticity That property of a material by virtue of which it tends to recover its original size and shape after deformation. If the strain is proportional to the applied stress, the material is said to exhibit Hookean or ideal elasticity.

Elastic Limit The greatest stress a material is capable of sustaining without any permanent strain remaining when the stress is released.

Elastomer A material which at room temperature can be stretched repeatedly to at least twice its original length and, upon release of the stress, will return with force to its approximate original length. Plastics with such or similar characteristics are known as elastomeric plastics. The expression is also used when referring to a rubber (natural or synthetic) reinforced plastic, as in "elastomer-modified" resins. May be either thermosetting or thermoplastic.

Electrical Relating to electricity but not dealing with its properties or characteristics.

Electrical Hold Value The minimum amount of current which will sufficiently energize a relay and maintain electrical contact. Also called Hold Current.

Electrical Isolation Two conductors isolated from each other electrically by a layer of insulation.

Electrically Hot Case A hybrid package whose case is part of the grounding circuit.

Electrical Resistance Test A test which measures the resistance in circuits to insure reliable connections.

Electric Field A region where a voltage potential exists. The potential level changes with distance and the strength of the field is expressed in volts per unit distance.

Electric Field Intensity The force exerted on a unit charge. The field intensity E is measured by

$$E = \frac{q}{4(3.1416) \, e \, r^2}$$

where r is the distance from the charge q in a medium having a dielectric constant e.

Electric Strength The maximum potential gradient that a material can withstand without rupture. The value obtained for the electric strength will depend on the thickness of the material and on the method and conditions of test. Also known as Dielectric Strength or Disruptive Gradient.

Electrode A conductor of the metallic class through which a current enters or leaves an electrolytic cell, at which there is a change from conduction by electrons to conduction by charged particles of matter, or vice versa.

Electroless Deposition The depositing of a conductive material from an autocatalytic plating solution without an electrical current flowing. Can be deposited on a dielectric material.

Electroless Plating The deposition of metallic particles from a chemical solution without an electrical current flowing. A very controlled and uniform thickness process.

Electrolytic Corrosion Corrosion caused by electrochemical erosion.

Electrolytic Plating Plating which is deposited from a plating solution by the application of an electrical current.

Electrolytic Tough Pitch A process or method used in preparing a type of raw copper which has excellent physical and electrical properties.

Electromagnet A coil of wire wound around an iron core which produces a strong magnetic field when the coil is energized.

Electromagnetic Field A rapidly moving electric field and its associated moving magnetic field, located at right angles both to the electric lines of force and to their direction of motion.

Electromagnetic Interference (EMI) A noise which may be conducted or radiated. Conducted EMI noise ranges from approximately 10 K Hz to 20 M Hz at 10 volts or less. Radiated EMI noise is usually in excess of 100 KHz.

Electromagnetic Pulse (EMP) An electromagnetic energy pulse from a nuclear blast or source.

Electromotive Force (EMF) The voltage or pressure that causes current to flow in a circuit.

Electron A negatively charge particle which orbits around the nucleus of an atom.

Electron Beam Bonding A process using a stream of electrons to heat and bond two conductors in a vacuum.

Electron-Beam Lithography A process in which radiation sensitive film or photo resist is placed in a vacuum chamber of a scanning beam electron microscope and exposed by a electron bean under digital computer control and subsequently developed by conventional processes.

Electron Beam Welding The process using a controlled electron beam focused on a small area; converting kinetic energy into high temperatures on impact with the part to be welded.

Electronic Hook-Up Wire Wires used to make connections between electrical components in electronic assemblies.

Electronic Interference An electrical or electromagnetic disturbance which causes undesirable responses in electronic equipment.

Electronic Packaging The combination of engineering and manufacturing technologies required to convert an electrical circuit into a manufactured assembly.

Electro-Optical See Optoelectronic. (Note: Optoelectronic is the preferred term.)

Electroplating The electrodeposition of an adherent metal layer on a conductive part or parts for environmental protection or decorative finish.

Electrostatic Discharge (ESD) The discharge of a static charge on a surface or body of a material or component through a conductive path to ground. An electronic component can be severely damaged if not property grounded to bleed off any static charge before it builds up to a high level charge. (DOD-HKBK-263 Transfer of electrostatic charge between bodies at different potentials caused by direct contact or induced by an electrostatic field).

Electrotinning The electroplating of tin on the surface of a part.

Element A part of an IC which contributes to its electrical characteristics. An active element exhibits gain such as a transistor; while a passive element does not have gain such as a capacitor or resistor.

Elongation The increase in gage length of a tension specimen, usually expressed as a percentage of the original gage length. (See Gage Length.)

Embed To encase completely a component or assembly in some material - a plastic, for instance. (See Cast and Pot.)

Emissivity The ratio of the radiant energy emitted by a source to the radiant energy of a perfect radiator (black box) having the equivalent surface area and same temperature and other conditions.

Emitter (1) In fiber optics, the source of optical power. Also called Source. (2) A region of the transistor from which charge carriers are injected into the base.

Emitter-Coupled Logic (ECL) A form of current mode logic in which the emitters of two transistors are connected to a single current carrying resistor in a way that only one transistor conducts at a time.

Emitter Electrode The metal pad which is connected to the emitter area of a transistor element.

Emulsion A light sensitive material used in the manufacture of printed circuits as well as to coat the mesh of screens in thick film technology.

Enameling A process which provides a glassy dielectric finish, which is virtually pore-free, on metal core substrates or wires.

Encapsulate To coat or embed a component or assembly with a conformal or thixotropic coating by dipping, brushing, spraying or potting.

End Bell An adaptor which attaches to the rear of a plug or receptacle.

End Of Life (EOL) The "wearing out" or the end of the useful operating life of a component or system and is expressed in a unit of time.

End Positioning Mounting A terminal block equipped with end section holes to accept screws without interfering with the contacts.

End Termination The metalized ends of discrete components, such as capacitors, or the metalized pads on passive chip devices which are used for making electrical contact.

Energy Of A Charge Measured in ergs (E), where the charge, Q, and the potential, V, are in electrostatic units. $W = 1/2\ QV$.

Energy Of The Electric Field Represented by

$$E = \frac{K\ H^2}{8(3.1416)}$$

Where K is the speciafic inductive capacity, H is the electric field intensity in electrostatic units and   the energy of the field, E, in ergs per cm^2.

Engaging And Separating Force The force required to engage and separate contact elements in mating connectors.

Engineering Change (EC) A change in the electrical or mechanical design or a material change.

Entity Two distinct groups of electrical circuits separated by a boundary and by input and output connections.

Entrapment The trapping of air, flux, and particulate in a medium which cannot escape.

Environment The combination of temperature, humidity, pressure, radiation, magnetic and electric fields, shock, and vibration which influence the performance of a device.

Environmental Seal A type of seal which keeps out moisture, air, and dust which might reduce the performance of a device.

Environmental Stress Cracking
The susceptibility of a thermoplastic article to crack or craze under the influence of certain chemicals or aging, or weather, and stress. Standard ASTM test methods that include requirements for environmental stress cracking are indexed in Index of ASTM Standards.

Environmental Test
A series of tests used to determine the external influences affecting the structural, mechanical, and functional integrity of an electronic package, assembly, or system.

Epitaxial
Pertaining to a single-crystal layer on a crystalline substrate and having the same crystalline orientation as the substrate.

Epitaxial Growth
A process in which layers of materials are grown on a selected substrate. Silicon is grown in a silicon substrate or those having compatible crystallographies.

Epitaxial Layer
A crystal layer with the same crystal orientation as the parent or base material.

Epoxy
Thermosetting polymers containing the oxirane group. Mostly made by reacting epichlorohydrin with a polyol like bisphenol A. Resins may be either liquid or solid.

Epoxy Glass
A mixture of glass fibers or woven glass cloth impregnated with an epoxy resin.

Epoxy Resins
Materials which form straight chain thermoplastic and thermosetting resins, and have excellent mechanical properties, dimensional stability, and are used extensively in electronic packaging.

Epoxy Smear
Epoxy resin which has been deposited on edges of copper in holes during drilling either as a uniform coating or as scattered patches. It is undesirable because it can electrically isolate the conductive layers from the plated-through-hole interconnections.

Ester
A class of organic compounds which are formed by the reaction of an acid with an alcohol with the elimination of water.

Etchant
A solution which is used to remove a conductive material, which is bonded to a base material, by chemical reactions.

Etchback
The controlled removal of all of the components of the base material by a chemical process acting on the sidewalls of plated-through holes to expose additional internal conductor areas.

Etched Metal Mask
A thin metal sheet on which a pattern is etched and is used in screening.

Etched Wiring Substrate
A printed conductive pattern formed by the chemical removal of conductive material that is bonded to a (dielectric) base material.

Etch Factor The ratio of depth of etch to the amount of undercut.

Eutectic A mixture whose melting point is lower than that of any other mixture of the same ingredients.

Eutectic Alloy An alloy having the same temperature for melting and solidus.

Exotherm The characteristic curve of resin during its cure, which shows heat of reaction (temperature) vs. time. Peak exotherm is the maximum temperature on this curve.

Exothermic A chemical reaction in which heat is given off.

Expanded Contact A contact made to the semiconductor die where the wire bonded to an area remote from the part to be electrically connected so that a lateral interconnection path for the current is required.

Expansion Connector A type of connector with a built-in flexible connection capability between a rigid conductor and the electrical equipment.

Exponential Failures Failures that occur at an exponentially increasing rate.

Exposure Subjecting photosensitive materials to radiant energy, such as light to photo resist to produce an image.

Extender An inert ingredient added to a resin formulation chiefly to increase its volume.

External Leads The flat ribbons or round wires which extend from an electronic package for input or output power, signals, or ground.

External Resistance A term used to represent thermal resistance from an external point on the outside of an electronic package to a point at ambient temperature.

Extraction Tool A device which is used for removing contacts from a connector.

Extrinsic Properties Properties in semiconductors which are caused by impurities in a crystal.

Extrinsic Semiconductor A semiconductor whose electrical properties are dependent on impurities.

Extrusion The compacting of a plastic material and the forcing of it through an orifice.

Eyelets Small metal tubes which are used as terminals after being inserted in a printed circuit board which provides mechanical support for component leads as well as reliable electrical connections.

Eyelet Tool In ribbon wire bonding, a tool with a square protuberance under the bonding tool surface which presses into the conductor and prevents slippage between the wire or conductor and tool interface.

Fabric A planar structure produced by interlacing yarns, fibers, or filaments.

Fabricate To work a material into a finished part by cutting, punching, drilling, tapping, machining, fastening and other operations.

Fabrication Tolerance The minimum and maximum variation in the characteristics of a part, when related to other defined variations will allow operation of the equipment within specified limits.

Face Bonding In face bonded semiconductor chip bonding; the circuitry side faces the substrate. Two common face bonding methods are Flip Chip and Beam Lead Bonding. The opposite of Back Bonding.

Face Seal A type of design which eliminates any voids at the interface of the plug and receptacle when they are engaged.

Fadeometer Equipment which is capable of accelerating the fading of resins and other materials by exposure to high intensity U.V. rays and determining the resistance of these materials to fading.

Failure A partial or total cessation of the function of a device attributed to electrical, physical, chemical, or electromagnetic damage and can be of temporary or permanent in nature.

Failure, Adhesive Rupture of an adhesive bond, such that the separation appears to be at the adhesive-adherend interface.

Failure Analysis A determination of the reason for the failure of a part to function at a specified level.

Failure Mechanism A physical or chemical defect that results in intermittent degradation or complete failure.

Failure Mode The manner in which a failure occurred such as the operating condition at the time of the failure.

Failure Rate The rate at which devices from a given number of devices can be expected to fail (or failed) as a function of time.

Fan In The maximum number of leads that can be connected to the output of a digital device.

Fan Out The maximum number of loads that can be connected to the input of a digital device.

Farad A unit of capacitance. When a capacitor is charged with one coulomb, it produces a difference of potential of one volt between its terminals.

Fatigue The weakening of a material caused by the application of stress over a period of time.

Fatigue Factor The factor or force which causes the failure of a material or device when placed under repeated stress.

Fatigue Tests Tests which require the application of a high stress and a low number of cycles or a low stress and a high number of cycles.

Fault-Closure Current Rating The rms fault current rating which a load break connector can engage under certain conditions.

Feedthrough A conductor which extends through the thickness of a substrate or printed wiring board thereby electrically connecting circuits on both surfaces. Also called an Interfacial Connection.

Feed Through Insulator A tubular shaped device made of a non conductive type material which is mounted on a metal chassis or bulkhead and is used to surround a conductor to prevent the conductor from shorting to ground.

FEP See Fluorinated Ethylene Propylene.

Ferrite A powdered, compressed, sintered, magnet material which has high resistivity. It consists of ferric oxide combined with one or more metals.

Ferroelectric Material A nonlinear dielectric material in which electric dipoles line up by mutual interaction. (e.g. barium titanate.)

Ferromagnetic Material A material having a high permeability and varies with the magnetizing force. (e.g. iron, cobalt, nickel and their alloys.)

Ferrule A short metal tube which is used to make a solderless electrical connection to a shielded or coaxial cable.

Fiber A relatively short length, small diameter threadlike material such as cellulose, wool, silk, or glass yarn.

Fiber Exposure A condition in which glass cloth fibers are exposed on machined or abraded areas.

Fiber Glass An individual filament made by attenuating molten glass. A continuous filament is a glass fiber of great or indefinite length; a staple fiber is a glass fiber of relatively short length (generally less than 17 in.).

Fiducial Mark A geometric configuration which is incorporated into the artwork of a PWB and used by a vision system to identify the artwork location and orientation.

Field Effect Transistor (FET) A semiconductor which is controlled by voltage. The current is controlled between the source terminal and drain terminal by voltage applied to a nonconducting gate terminal.

Field Replaceable Unit (FRU) An electrical subsystem which can easily and readily be replaced in the field or area of operation.

Field Trimming The trimming of a resistor to a specific output voltage or current. (Also known as Functional Trimming.)

Filament A single fiber of indefinite length.

Filament Winding A process for fabricating a composite structure in which continuous reinforcements (filament, wire, yarn, tape, or other) either previously impregnated with a matrix material or impregnated during the winding are placed over a rotating and removable form or mandrel in a previously prescribed way to meet certain stress conditions. Generally the shape is a surface of revolution and may or may not include end closures. When the right number of layers are applied, the wound form is cured and the mandrel removed.

Filled Plastic A plastic material to which has been added ceramic, silica, or metal powders to improved a specific property such as thermal conductivity, or to lower its cost.

Filler A material, usually inert, that is added to plastics to reduce cost or modify physical properties.

Fillet A concave junction formed at the intersection of two surfaces.

Film (1) A thin sheet or coating having a nominal thickness not greater than 0.010 inches. (2) A thin coating or layer of material. Thin films are deposited by vacuum evaporation, sputtering, and plating while thick films are screen printed.

Film Adhesive A class of adhesives provided in dry-film form with or without reinforcing fabric and cured by heat and pressure.

Film Conductor An electrically conductive material deposited on a substrate by screen printing, plating, or by vacuum deposition.

Film Integrated Circuit A circuit, in film form, made up of elements and formed in place on a substrate. Also called a Film Microcircuit and can be made by thin or thick film techniques.

Film Microcircuit An electrical network forming electrical interconnections to various devices by thin or thick film techniques.

Film Network An electrical network consisting of thin or thick film devices on a substrate.

Film Resistor A film type device, made of resistive material, on a substrate.

Filters A fibrous, organic material which is used to remove solids and organic impurities from liquid solutions.

Final Seal In microelectronic packages, the final manufacturing operation that encloses or seals the package so that no further internal processes can be performed without delidding.

Fine Leak A very small leak in a sealed package having a differential air pressure of less than 10^{-5} cm^3/sec at one atmosphere.

Fine Pitch Surface mount devices with a lead pitch of less than 0.025 inches.

Fingers See edgeboard Contacts.

Fire In thick film technology, the heating of screened circuits, resistors, and capacitors to elevated temperatures to achieve specific properties.

Firing Sensitivity The percent of change in the film characteristics during firing due to a change in peak firing temperature. It is expressed in units of percentage/degree Celsius.

First Article One of the first parts manufactured which is used for inspection and test purposes, to assure it meets the requirements, prior to manufacturing a larger number of parts.

First Bond The initial bond in a series of two or more bonds to form a conductive connection.

First Radius The radius of the front edge of a bonding tool foot.

First Search The part of the cycle in which the final adjustment are made to the machine in the location of the first bonding area under the tool prior to lowering the tool to make the first bond.

Fish Paper Electrical-insulation grade of vulcanized fiber in thin cross section.

Fissuring The cracking of dielectric or conductor materials during firing.

Fixed Contact A contact which is permanently fixed/attached in the insert during the molding process.

Flag Terminal A type of terminal in which the tongue extends from the side of the terminal instead of the end of the barrel.

Flame Off See Burn Off.

Flame Resistance The property of a material to extinguish the flame after the source of heat is removed.

Flame Retardant Resin A resin material which is compounded with certain chemicals to reduce or eliminate its tendency to burn.

Flame Retarder A material, which when added to a resin mixture will result in the self extinguishing of the flame after the source of fire has been removed.

Flammability The measure of a material's ability to support combustion.

Flange That part of a connector which is elliptical in shape and extends outside or beyond the periphery of the connector. Its sole purpose is to provide holes to accommodate a screw or bolt for mounting to a panel or the mating connector half.

Flanged Spade Tongue Terminal A terminal with a slotted tongue and with the ends of the tongue bent up or down to prevent the terminal from disengaging from its captive hardware.

Flash Extra plastic attached to a molding along the parting line; under most conditions it would be objectionable and must be removed before the parts are acceptable.

Flashover An electric discharge along the surface of a dielectric material caused by an overvoltage.

Flash Plating A very thin deposition of electroplated material, such as copper or nickel, on the part to be plated for subsequent thicker plating.

Flat Base A terminal block equipped with top to bottom feedthroughs, but without circuit isolation on the bottom side.

Flat Cable A multiconductor cable assembly whose conductors are laid out in the same plane and whose top and bottom surfaces are flat.

Flat Cable Connector A connector specifically designed to accommodate and terminate flat cable. Equipped to handle flat conductors, flat cable, or flat cable with round conductors.

Flat Pack A package with its leads extending from the sides and parallel to the base and containing an integrated circuit.

Flex Damage The damage caused by a sharp bend in the cord as it enters the housing. To relieve this problem a flex relief is added to restrict the flexing concentration and cause the cord to bend in a wider arc.

Flexibilizer A material that is added to rigid plastics to make them resilient or flexible. Flexibilizers can be either inert or a reactive part of the chemical reaction. Also called a Plasticizer in some cases.

Flexible Circuit Carrier Thin flexible materials, such as Kapton or polyimide, on which printed circuits have been deposited by pattern plating.

Flexible Coating A coating which is still flexible after curing.

Flexible Printed Circuit An arrangement of printed wiring utilizing nonrigid base material for cover and insulating layers.

Flexural Modulus The ratio, within the elastic limit, of stress to corresponding strain. It is calculated by drawing a tangent to the steepest initial straight-line portion of the load-deformation curve and calculating by the following equation:

$$E_b = \frac{L^3\, m}{4\, b\, d^3}$$

where E_b = modulus

    b = width of beam tested
    d = depth of beam
    m = slope of the tangent
    L = span, in.

Flexural Strength The strength of a material in bending expressed as the tensile stress of the outermost fibers of a bent test sample at the instant of failure.

Flip Chip A semiconductor die having all terminations on one side in the form of solder pads or bump contacts. After the surface of the chip has been passivated, it is flipped over for attachment to a matching substrate.

Flip Chip Mounting A technique of mounting flip chips on thick or thin film circuits without subsequent wire bonding.

Floating Bushing An eyelet type bushing located in the plug mounting holes which allows complete freedom of motion in all directions during engagement of the plug and receptacle. Primarily used for easy alignment.

Floating Ground (1) An electrical ground that is not connected to earth. (2) A ground circuit that does not allow a connection between the power and signal ground for the same signal.

Flood Bar A screen printing device which drags the paste back to the starting point after the squeegee has printed without pushing it through the mesh of the screen.

Floor Planning A layout of the approximate placement and orientation of logic and memory circuitry groupings in a package prior to final design.

Flow Regime, Laminar The smooth flow of undisturbed fluid layers.

Flow Regime, Turbulent The irregular and fluctuating flow of fluid particles.

Flow Soldering A soldering method which involves the immersion of the metallized areas to be soldered in a molten solder bath to produce an acceptable solder joint.

Fluorinated Ethylene Propylene (FEP) A copolymer of tetrafluoroethylene and hexafluro- propylene, such as DuPont's Teflon.

Fluorocarbon An organic compound having fluorine atoms in its chemical structure. This property usually lends chemical and thermal stability to plastics.

Flush Conductor A conductor whose longitudinal surface is in the same plane as the adjacent insulating material.

Flush Mount A type of device whose body is recessed in a panel or chassis and whose face is even with or projects slightly above the surface of the panel or chassis.

Flux A chemical which attacks and removes oxides from the surface of metals so that the molten solder can wet the surface to be soldered.

Flux, Activity The cleaning and wetting of a metal surface, during heating, caused by the flux. The flux activity increases continuously, during heating, and decreases above a temperature because of degradation of the active substances.

Flux, Inorganic Acid (IA) Typically hydrochloric, hydrofluoric, or orthophosphoric acid.

Flux, Organic Acid (OA) Typically lactic, oleic, stearic, glutamic, or phthalic acid dissolved in water, organic solvent, petrolatum paste, or polyethylene glycol. Also known as a Organic Water Soluble Flux.

Flux Residue Particles of flux remaining after soldering and cleaning.

Flux, Rosin (R) An organic acid (primarily abietic acid) which is a natural organic resin derived from pine tree sap. The rosin is dissolved in isopropyl alcohol, organic solvent or polyethylene glycol. It is the least active rosin flux and the residue is a hard transparent film with good electrical insulation properties and resistant to water absorption.

Flux, Rosin Activated (RA) Same as type R but with an even stronger activator than is type RMA.

Flux, Rosin Mildly Activated (RMA) Same as type R but with the additional of a mild activator. Most commonly used in electrical soldering applications.

Flux, Synthesized Activity (SA) Contains organic derivatives of sulphur and phosphorous acids which are soluble in fluorocarbon solvent blends. Possesses high flux activity which is corrosive and requires washing.

Foam Fluxing A type of flux in the form of foam. It is commonly used in wave soldering wherein the flux is transformed from a liquid flux to a foam by a porous diffuser.

Follower A tube or sleeve which compresses the grommet thereby improving or tightening the seal around the wire as it enters the connector.

Foot Length The long dimension of the bonding surface of a wedge type bonding tool.

Footprint The area on a substrate intended for the mounting of a particular component. Conforms to the geometric pattern of a chip.

Forming Gas Usually nitrogen gas, with small amounts of hydrogen or helium added to the nitrogen that is used as a blanket to cover a part to prevent oxidation of metal surfaces.

Four Harness Satin This fabric is also named crowfoot satin because the weaving pattern when laid out on cloth design paper resembles the imprint of a crow's foot. In this type of weave there is a 3 by 1 interlacing. That is, a filling thread floats over the three warp threads and then under one. This type of fabric looks different on one side than on the other. Fabrics with this weave are more pliable than either the plain or basket weave and, consequently, are easier to form around curves.

FR-4 A fire retardant epoxy resin glass cloth laminate as designated by the National Electrical Manufactures Association, NEMA.

Frame The outer most portion of a connector which has a removable body or insert. It is usually made of metal and supports the insert and used for mounting the connector to a panel or the mating connector half.

Frequency The number of complete cycles per unit of time.

Fresnel Reflection Losses In fiber optics, it is the losses which occur at the terminus interface due to differences in refractive indexes.

Fretting A method of maintaining good electrical surfaces of mating contacts by the movement of parts thereby exposing fresh metal.

Frit Finely ground glass used in thick film pastes which melts during firing and provides adhesion to the substrates for powdered metals and metal oxides.

From-To List A set of wiring instructions which lists the terminals to be connected.

Front-End-Of-The-Line (FEOL) The first part of the fabrication of devices such as transistors and resistors. The first part includes all the processes used in the manufacture of the IC devices.

Front Mounted A connector or device which is installed or removed from the outside of a panel or chassis.

Front Release Contacts A type of connector in which the contacts are engaged and released from the front side with the aide of tool and subsequently pushed out of the back of the connector.

Full-Cycling Control A type of crimping tool which cannot be opened until after the crimping has reached its maximum extent.

Functional Test An electrical test in which an assembly is subjected to actual operating conditions.

Funnel Entry A terminal or connector whose end is flared or funnel shaped to allow easier insertion of stranded wire into the wire barrel.

Furnace Active Zone The portion of a multizoned muffle furnace that is thermostatically controlled.

Furnace Profile A graph of time versus temperature, or a position in a continuous thick film furnace versus temperature.

Furnace Slave Zone A section of a multizoned muffle furnace where power applied to the heating element is a set percentage of the power supplied to the active zone.

Fuse Terminal Block A terminal block which is equipped with a fuse.

Fusing (1) The melting and cooling of two or more powder materials together so that they bond together in a homogeneous mass. (2) The process of heating and melting of a thin metal strip and subsequent resolidification.

G-10 A grade of epoxy impregnated glass cloth printed circuit board material which meets the National Electrical Manufacturers Association requirements. Has been largely replaced with the flame retardant equivalent NEMA Grade FR-4.

Gage Length The original of that portion of the specimen over which strain is measured.

Galvanic Displacement A chemical deposition of a very thin metallic coating on certain base metals by partial displacement. (Sometimes called Immersion Plating.)

Gang Bonding A process in which several mechanical or electrical bonds are made by a single stroke of a bonding tool.

Gang Disconnect A connector which can rapidly connect or disconnect many circuits at one time.

Gas Blanket An inert gaseous atmosphere such as nitrogen flowing over heated parts to prevent oxidation.

Gas Tightness A device which is capable of resisting the flow of harmful gases which can cause corrosion.

Gate (1) A circuit with one output and many inputs, designed so that the output is energized only when a specific combination of pulses is present at the inputs. (2) In injection and transfer molding, the orifice through which the melt enters the cavity.

Gate Array A geometric configuration of basic gates in a chip. Some arrays contain literally hundreds and thousands of gates depending on their complexities.

Gate, Logic A circuit which combines two input signals into one output signal in accordance to the function it performs.

Gate, Sea Of A gate array in which the circuitry required for interconnecting the fixed array of logic cells are in rows and columns and from one logic unit to another.

Gate, Structural A frame on which several printed wiring assemblies are mounted and attached to a chassis by a hinge. It can easily be opened for access and servicing.

"G" Dimension The distance between two crimped points of a connector. Also called the "T" Dimension.

Gel The soft rubbery mass that is formed as a thermosetting resin goes from a fluid to an infusible solid. This is an intermediate state in a curing reaction, and a stage in which the resin is mechanically very weak. Gel point is defined as the point at which gelation begins.

G

Gel Coat A resin applied to the surface of a mold and gelled prior to lay-up. The gel coat becomes an integral part of the finished laminate, and is usually used to improve surface appearance, etc.

Germanium (Ge) A chemical element, atomic number of 32, grayish white in color, with semiconductor properties. Used in transistors and crystal diodes.

Giga A prefix which means it is to be multiplied by one billion (1,000,000,000).

Glass An inorganic, non crystalline, non metallic material made by melting silicon, soda ash, lime and other similar materials.

Glass Binder A glass powder which is added to resistor and conductor pastes to bind the particles together after firing.

Glass - Ceramic A solid, nonmetallic material having a fine grain, nonporous microstructure formed by the controlled crystallization of glasses.

Glass + Ceramic A solid, nonmetallic material made from mixing crystalline ceramic and glass and subsequently sintering into a composite material.

Glass +/- Ceramic Materials made from glass - ceramic and glass + ceramic.

Glass Fabric A woven cloth made from glass yarns. The glass yarns are made from glass filaments.

Glassivation A protective layer of glass which is applied to die surfaces. Usually silicon dioxide or silicon nitride. Also called Passivation.

Glass Phase That part of the firing cycle when the glass binder is in the molten state.

Glass Preforms A cylindrical shaped, hollow, glass configuration which is either opaque or clear. When pressed into shape with a binder and sintered, forms a hermetic seal in metal packages.

Glass To Metal Seal A hermetic seal between glass and metal parts. The seal is made by fusing together glass and metal alloys having nearly the same coefficients of thermal expansion.

Glass Transition Point Temperature at which a material loses its glasslike properties and becomes a semiliquid.

Glass Transition Temperature (Tg) The temperature at which plastics and base laminates, for example, soften and begin to expand independently of the glass fabric expansion rate.

Glazed Substrate A glass coating on a ceramic substrate which provides a smooth and nonporous surface.

Glazing The application of a glass coating to a ceramic or metal surface, and subsequent firing, which provides a smooth surface and seals the surface against water absorption.

Global Wiring The interconnecting of components which are mounted on a package, not inside the package.

Glob Top The application of a glob of encapsulation material such as an epoxy or silicone, to the top of a chip after electrical test, wire bond, inspection, etc. to completely cover the entire chip.

Globule Test A test which measures the wettability of a component lead, using a globule of solder, against a known standard of time.

Glossy A smooth, shiny, nonporous surface formed by the glass matrix in a conductor or resistor paste.

Glue Line The layer of adhesive which attaches two adherends. (Also called Bond Line).

Glue-Line Thickness Thickness of the fully dried adhesive layer.

Glycol An alcohol containing two hydroxyl (-OH) groups.

Gold Flash A very thin layer of gold, approximately 0.0001 inches or less.

Gradient The rate at which a variable quantity increases or decreases.

Grain Growth The growing in size of the crystal grains in glass or metals over a period of time.

Gram-Force A unit of force required to support a mass of one gram. One gravity unit of acceleration X's one gram of mass = one gram-force.

Graphite Fibers High-strength, high-modulus fibers made by controlled carbonization and graphitization of organic fibers, usually rayon, acrylonitrile, or pitch.

Green Tape An unfired flexible ceramic material having a thin, uniform thickness.

Grid A network of equally spaced lines forming squares.

Grid Spaced The positioning of contacts or pins of a connector in a rectangular configuration such that they are equally spaced from each other.

Grommet A plug shaped device, usually made of an elastomeric or rubber material, used to seal the rear portion of a connector against moisture, dirt, air, and other harmful contaminants.

Groove The slotted area in a connector which exerts pressure on the cable.

Gross Leak A leak in a sealed electronic package greater than $10cm^{-5}/sec^3$. at one atmosphere of differential air pressure.

Grounded Parts Parts which are electrically connected and are at the same potential as the earth.

Ground Plane A conductive layer on a substrate or printed wiring board which serves as a common circuit return, EMI/RFI shielding, or heat sinking.

Guide Pin A hardened or wear resistant metal pin which extends beyond one mating surface of a two piece connector which assures the accurate alignment of the contacts during the engagement of the connector.

Gull Wing The shape of the leads of surface mounted devices conforming to shape of the wings of a sea gull. They are bent outward, downward, and then upward from the body of the package. This provides feet for mounting and soldering with inherent mechanical compliance.

Hairpin Mounting A high density type of mounting axial components in which one lead is bent to form a 180 degree bend or hairpin configuration and the other lead is left straight and the component is mounted vertically. Not considered reliable in vibration, shock, and impact environments.

Hall Effect The development of a potential difference between the two edges of a strip of metal in which an electric current is flowing longitudinally, when the plane of the strip is perpendicular to a magnetic field.

Halo Effect The existence of a glass halo around certain thick film conductors. It can be avoided by changing furnace profiles or types of materials.

Halogenated Hydrocarbon Solvents Organic solvents containing elements of chlorine, fluorine, bromine, and iodine.

Halogenated Polyester A polyester resin which has been modified with halogens which reduces its flammability.

Haloing A light area around holes or other machined areas on or below the surface of the base laminate and is the result of fracturing or delaminating of the base material.

Hand Lay-Up The process of placing (and working) successive plies of reinforcing material or resin-impregnated reinforcement in position on a mold by hand.

Hand Soldering A process in which solder connections are formed by using a hand-held soldering iron for the application of heat.

Hardened Circuit A circuit which can resist damage from transient over-voltages or other forces such as radiation hardening.

Hardener A chemical added to a thermosetting resin for the purpose of causing curing or hardening. Amines and acid anhydrides are hardeners for epoxy resins. Such hardeners are a part of the chemical reaction and a part of the chemical composition of the cured resin. The terms hardener and curing agent are used interchangeably. Note that these are different from catalysts, promoters, and accelerators, which are not a part of the chemical composition of the cured resin.

Hard Glass A type of glass which has a softening temperature greater than 700 degree Celsius such as the borosilicate glasses.

Hardness See Indentation Hardness.

Hard Solder Those solders which have melting points above 425 degree Celsius.

Hardware Components or devices such as pins, clamps, screws, terminals, etc.

Harness A bundle of wires, with or without breakouts, which are held together with ties or covered with a braided or plastic sheath.

Header The base of a package from which the external leads extends. It does not include the lid.

Heat Aging An environmental test used to indicate the relative resistance to heat degradation.

Heat Clean The process of removing organic materials from glass cloth by heating to approximately 650 degrees - 700 degrees Fahrenheit for periods up to 50 hours.

Heat Column The heating element in a eutectic die bonder or a wire bonder. Its purpose is to heat the substrate to bonding temperatures.

Heat-Distortion Point The temperature at which a standard test bar deflects 0.010 inches under a stated load of either 66 or 264 lb/in^2. See Standard Method of Test for Deflection Temperature of Plastics under Load, ASTM D 648.

Heat Endurance The length of time and the degree of temperature rise a material or component can withstand before failing a physical or mechanical requirement.

Heat Flux The flow of heat away from a heat source.

Heat Resistance The ability of a material to retain its mechanical and electrical properties as measured during exposure of the material over a temperature range and for a period of time.

Heat Sealing A method of joining plastic films by simultaneous application of heat and pressure to areas in contact. Heat may be supplied conductively or dielectrically.

Heat Shrinkable A term which describes sleeves, tubes, boots, tape, and other forms of plastic which shrink when heated to encapsulate, insulate, and protect connections, splices, and terminations.

Heat Sink Any device that absorbs or transfers heat from the generating source.

Heat Soak The heating of an assembly over a period of time to allow all the components, circuits, packages, etc. to stabilize in temperature.

Heel Break The fracture or break of the lead at the heel of the wire bond.

Heel Crack A crack across the wire bond width located in the heel area.

Heel of the bond The part of the lead adjacent to the bond that is deformed by the edge of the bonding tool in making a wire bond.

Helium The type of gas used in the detection of fine leaks. It is used because of its small size molecules and is not flammable.

Helium Leak Check A method for detecting fine leaks in hermetically sealed packages in which the trace gas is helium. The leak rate is expressed in terms of cubic centimeters of helium per second at a pressure differential of one atmosphere.

Hermaphroditic, Connector And Contact Parts which are identical at their mating surfaces.

Hermetic Permanently sealed by fusion or soldering such as metal to metal or metal to glass to prevent the transmission of moisture, air, and other gases. Typical maximum leak rates of hermetically sealed packages are 1×10^{-8} c.c. of helium per second at a pressure differential of one atmosphere.

Hertz A unit of frequency equal to one cycle per second. Abbreviation is Hz.

Hidden Via See Buried Via.

High Altitude Electromagnetic Pulse (HEMP) An energy pulse which is caused on or near the surface of the earth by the radiated electromagnetic field from a nuclear blast above the atmosphere.

High Density Pertains to the number of parts and interconnections in an electronic package. High density packaging implies fine feature sizes, very close feature spacing and high input - output (I/O) or pin counts.

High Energy Surface Materials which have surface free energies in the range of 5,000 - 500 ergs/cm^2. Normally they have a high melting point such as metals, metal oxides, nitrides and glasses.

High Frequency In the radio frequency spectrum, the band from 3 to 30 MHz.

High-Frequency Preheating The plastic to be heated forms the dielectric of a condenser to which is applied a high-frequency (20 to 80 MHz) voltage. Dielectric loss in the material is the basis. The process is used for sealing vinyl films and preheating thermoset molding compounds.

Hi-K Abbreviation for High Dielectric Constant.

High-K Ceramic A ceramic dielectric composition, such as barium titanate, which exhibits a high dielectric constant, and nonlinear voltage and temperature response.

High Level Noise Tolerance (NTH) The noise tolerance level of the receiver when the input signal is in its UP state.

High-Purity Alumina Alumina having a purity in excess of 99% of Al_2O_3.

High Speed Circuitry Circuitry having a clock rate in the range of approximately 8 - 300 MHz.

High Temperature Stability The ability of a material to sustain high temperatures without any changes in its properties caused solely by heat.

High Voltage (1) An ac circuit whose difference in potential is at least 1 kV rms. (2) A dc circuit which functions at a difference of at least 1.4 KV.

Hipot An electrical test which measures the voltage breakdown of a dielectric material. Also called High Potential.

Hi Rel High reliability. A device or component designed and manufactured to close tolerances, controlled process, and inspection requirements to provide a long service life.

Hold Current See Electrical Hold Value.

Holding Strength The ability of a connector to remain assembled to a cable or to the individual wires when under tension.

Homogeneous A material whose composition is uniform throughout.

Hood An enclosure which is attached to the rear of a connector which contains and protects wires and cable attached to the terminals of a connector. A cable clamp is considered part of the hood.

Hook Terminal A terminal with a hook shaped tongue.

Hook Tongue A type of terminal with a tongue which opens from the side rather than from the end.

Horn A cone shaped member which transmits ultrasonic energy from the transducer to the bonding tool.

Hostile Environment An environment which degrades electronic circuits and devices such as temperature excursions, humidity, radiation, etc.

Hostile Environment Connectors Connectors which are designed and engineered for operation in temperatures ranging from absolute zero to 675 degree Celsius and water tight conditions.

Hot Bar Soldering A process in which a heated bar solders all the leads of a device to the pads of a printed wiring board or substrate simultaneously.

Hot-Gas Reflow A process in which a heated gas is used to reflow solder and form solder joints and interconnections.

Hot-Line Clamp A type of connector which may be installed or disconnected by using an insulated stick or device while the conductor is energized. Also called a Live-Line Connector.

Hot-Melt Adhesive A thermoplastic adhesive compound, usually solid at room temperature, which is heated to a fluid state for application.

Hot Spot A point of maximum temperature on a component that is unable to dissipate heat into the surrounding areas.

Hot Tin Dipping A technique in which a coating of solder is applied to a metal part (a component lead, printed wiring board, etc.) by immersing the part in a molten solder bath.

Hot Zone That part of continuous furnace or kiln which is at maximum temperature and located between the preheat and cooling zones.

Housing The shell portion of a connector minus the insert, but including the positioning hardware.

Human Engineering The science which deals with the designing, building, and testing of mechanical equipment to meet the anthropometric, physiological, or psychological requirements of the people who use them.

Humidity The amount or degree of moisture in the air.

Humidity Resistant The ability of a material to resist the absorption of moisture.

Hybrid A mixture of two or more different technologies and components. Examples are: thin and thick film circuits and active and passive devices, etc.

Hybrid Circuit A circuit in which the active and passive components, which are manufactured separately, and are subsequently interconnected by various bonding processes using thin and thick film technologies.

Hybrid Electronic Assembly An electronic assembly containing discrete integrated circuit devices and other chip type components, thin or thick film circuits, and wire bonding or other techniques for interconnecting purposes.

Hybrid Electronics A technology utilizing thin and thick film circuitry, discrete integrated circuit devices, and other chip type components, with wire bonding or other techniques for interconnecting purposes to form electronic circuits.

Hybrid Group Hybrid microcircuits designed to perform the same types of basic circuit functions for the same supply, bias, and voltages.

Hybrid Integrated Circuit A microcircuit made up of thin film or thick film circuits on a substrate to which are mounted and bonded active and passive microdevices.

Hybrid Microcircuit Used interchangeably with Hybrid Integrated Circuit.

Hybrid Microelectronics The electronic art which applies to electronic systems using hybrid circuit technology.

Hybrid Microwave Circuit See Microwave Integrated Circuit.

Hybrid Module A special enclosure containing hermetically sealed hybrid packages, discrete passive components, and non hermetic hybrid modules which are conformably coated, all of which are electrically interconnected.

Hydrocarbon An organic compound having hydrogen and carbon atoms in its chemical structure. Most organic compounds are hydrocarbons. Aliphatic hydrocarbons are straight-chained hydrocarbons, and aromatic hydrocarbons are ringed structures based on the benzene ring. Methyl alcohol, trichloroethylene, etc., are aliphatic; while benzene, toluene, etc., are aromatic.

Hydrolysis Chemical decomposition of a substance involving the addition of water.

Hydrophilic Having an affinity for water; wettable.

Hydrophobic Having no affinity for water; nonwettable, tends to repel water.

Hydroxyl Group A chemical group consisting of one hydrogen atom plus one oxygen atom.

Hygroscopic The ability of a material to absorb and retain moisture from the atmosphere.

Hysteresis An effect in which the magnitude of a resulting quantity is different during increases in the magnitude of the cause than during decreases due to internal friction in a substance and accompanied by the production of heat within the substance. Electric hysteresis occurs when a dielectric material is subjected to a varying electric field as in a capacitor in an alternating-current circuit.

Icicle A term used to describe a solder projection.

Identification Pin A method of denoting pin 1 in a plug-in package. Usually done by coloring the glass bead or a die stamped number on the header or a square corner.

Imbedded Layer The conductor layer located between the insulating layers.

Immunity The characteristic of a system to be immune to disturbances.

Impact Resistance The resistance of a plastic material to fracture under shock or impact.

Impact Strength A measure of toughness of a material by the energy required to fracture a specimen with a single blow.

Impedance The total opposition that a circuit offers to the flow of alternating current or any other varying current at a particular frequency. It is a combination of resistance R and reactance X, measured in ohms and designated by Z.

$$Z = (R^2 + X^2)^{1/2}$$

Impedance Match When the impedance of a component, circuit or load is equivalent to the internal impedance of the source, or to the surge impedance of a transmission line.

Impregnate To force resin into every interstice of a part. Cloths are impregnated for laminating, and tightly wound coils are impregnated with liquid resin using air pressure or vacuum as the impregnating force.

Impulse A pulse of voltage or current in an infinitesimal time. (See Pulse.)

Impulse Ratio The ratio of a flashover or breakdown voltage of an impulse to the crest value of the power frequency flashover or breakdown voltage.

Inactive Flux A flux which becomes nonconductive after reaching soldering temperatures.

In Circuit Test An electrical test which is performed on individual components even though they are soldered in place.

Inclined Plane Furnace A furnace whose hearth is inclined such that a draft of oxidizing atmosphere will flow through the heated zones by natural convection.

Inclusion A foreign particle, metallic or nonmetallic, in a conductive layer, plating, or the base material.

Incomplete Bond A bond impression, less than normal size, because part of the bond is missing.

Indentation Hardness Hardness evaluated from measurements of area or indentation depth caused by pressing a specified tool or indentor into the material surface with a specified force.

Indentor A part of the crimping die set. It shapes the terminal barrel to the desired configuration while the nest provides the location and support of the crimping process.

Index Matching Materials In fiber optics it is those materials which are placed between the ends of the optical conductors to reduce the coupling losses.

Index Of Refraction The ratio of the velocity of a light wave in a vacuum to that in a specified medium.

Indirect Emulsion An emulsion which is transferred to the screen surface from a plastic carrier or backing material.

Inductance The property of a circuit or circuit element that opposes a change in current flow which causes current changes to lag behind voltage changes. Measured in henrys.

Induction Soldering A method of soldering in which the solder is reflowed when the work is moved slowly through an electromagnetic field.

Inductive Crosstalk Crosstalk which is created by the coupling of the electromagnetic field of one conductor on another.

Inert Atmosphere A gas atmosphere such as helium, nitrogen, or a mixture of gases that is nonoxidizing or nonreducing in the treatment of metals.

Inflammable Flammable, combustible, capable of catching fire, having a low ignition point.

Infrared (IR) A part of the electromagnetic spectrum, lying between wavelengths of 0.75 and 1000 micrometers.

Infrared Reflow A process in which an infrared light source is used to heat solder joints to their melting point. This process is usually performed in an IR furnace. PWB's with surface mounted devices and other types of packages and components, properly positioned, would be applicable to this process.

Inhibitor A chemical added to resins to slow down the curing reaction. Inhibitors are normally added to prolong the storage life of thermosetting resins.

Initiator A substance used to start a chemical reaction and is no longer needed once the reaction has started.

Injection Molded Card (IMC) A printed circuit board assembly which has been injection molded. These assemblies are not repairable.

Injection Molding A molding procedure whereby a heat-softened plastic material is forced from a cylinder into a cavity which gives the article the desired shape. Used with all thermoplastic and some thermosetting materials.

Ink A screenprintable thick film material or "paste" consisting of glass frit, metals, or metal oxides, and solvents.

Inner-Lead Bonding (ILB) Connections which are made between the chip and the etched conductors on the tape. This is usually a gang-bond type of connection.

Inorganic Chemicals Chemicals whose chemical structure is based on atoms other than the carbon atom.

Inorganic Pigments Natural or synthetic salts, which are carbon free and used for coloring plastic. Noted for their color stability and weather resistance.

In Process A point in the various manufacturing operations but before final testing is complete.

Input A signal which is applied to a circuit.

Input/Output (I/O) Referring to devices which are used to interface with a computer and the data to be received or transmitted.

Insert That part of a connector which holds the contacts in their proper arrangement and insulates them from each other and from the shell.

Insert Arrangement The number, spacing, and arrangement of contacts in a connector.

Insert Cavity A specific hole in the connector insert into which the contacts are placed.

Insertion The placing of the wire terminations of components into holes of a printed wiring board.

Insertion Force The force required to engage mating components usually measured in ounces.

Insertion Loss The difference between the power received at the load before and after the insertion of a component, connector, or device at some point in the line.

Insertion Mount Component (IMC) A leaded component specifically designed for automatic or semiautomatic mounting on PWB's. The leads are inserted in the holes of the board, to a controlled length and the component body to a prescribed height above the surface of the PWB and subsequently soldered to the circuitry on the board surfaces.

Insertion Tool A small hand tool which is used to insert contacts into the connector.

Insert Retention The force which must be applied in either direction (push or pull) that an insert is required to withstand without being removed from the connector shell.

Inspection Hole A hole located at one end of a barrel to allow visual inspection to assure the conductor has been inserted to the proper depth in the barrel before crimping.

Insulated Terminal A solderless terminal having an insulated sleeve over the barrel to prevent a short circuit during installation.

Insulation A nonconductive material with a high resistance to the flow of electric current.

Insulation Crimp The area of a terminal, splice, or contact that is formed around the insulation of a wire.

Insulation Displacement Connector A connector capable of making many electrical contacts through the insulation on a flat cable at one time by insulation piercing.

Insulation Grip Certain crimp-type contacts which have extended cylinders at the rear to accept the base wire and a short length of its insulation. The wire and insulation are held firmly in place after crimping.

Insulation Piercing A crimping method in which lances pierce the wire insulation and make contact with the conductor stands without stripping the insulation off the wire.

Insulation Piercing Terminal A crimping device which pierces wire insulation and makes electrical contact without stripping the insulation from the wires.

Insulation Resistance (1) The electrical resistance of the insulating material between any pair of contacts, conductors, or grounding devices in various combinations. (2) The ratio of the applied voltage to the total current between two electrodes in contact with a specific insulator.

Insulation System The sum total of the insulation materials which are used in a system to electrical insulate all electronic components.

Insulator A nonconducting material with a high resistivity used to support or separate conductors. Made from materials with a resistivity greater than 10^5 ohms/cm.

Insulator Metal Substrate Technology (IMST) A substrate having a metal core and coated with a dielectric such as porcelain or an epoxy and etched circuitry on one side.

Integrated Circuit (IC) A small chip of semiconductor material containing an array of active and/or passive components which are interconnected to form a functioning circuit.

Integrated Injection Logic Integrated circuit logic which uses bipolar transistor gates.

Interchip Wiring The conducting paths connecting circuits on one chip with those on other chips in order to complete the electrical circuits.

Interconnection The conductive path required to achieve connection from a circuit element to the rest of the circuit.

Interface The junction point or surface between two different or same materials.

Interface Connection A conductor which connects circuits on opposite sides of a printed wiring board. Also called a Feed-Through or Plated-Through-Hole.

Interfacial Bond An electrical connection between the circuitry on the two sides of a substrate.

Interfacial Connection The method used to connect circuitry on opposite sides of a printed wiring board such as a plated-through-hole, eyelet, etc.

Interfacial Gap The distance between the faces of a mated connector.

Interfacial Junction The meeting of the two faces of a connector.

Interfacial Seal A seal which is formed around each contact when the two mating halves of a connector are in compression. This is accomplished by the use of an elastomeric material which covers the entire interface area of the mating connective halves.

Interference Protection The methods/measures which provide shielding to sensitive areas of electrical equipment from EM/RF interferences.

Interlayer Connection An electrical connection between circuitry in different layers of a multilayer printed wiring board. Also called a Through Connection.

Intermateable Referring to connectors; they are connector halves which are made by various manufacturers which are interchangeable.

Intermetallic Bond The ohmic contact which is established when two metal conductors are welded or fused together.

Internal Resistance The thermal resistance encountered between the junction of two dissimilar materials inside a package to a point on the outside surface of the package.

Interpenetrating Polymer Network (INP) A resulting polymer mixture of a reaction between two or more crosslinked polymers other than the normal covalent bonding as in a copolymer.

Interstice A very small space between two objects which are purposely placed in close proximity.

Intraconnection The joining of conductors in a circuit on the same substrate.

Intrinsic Dielectric Strength The characteristic dielectric strength of a material.

Invar A nickel-iron alloy having a very low coefficient of thermal expansion.

I/O Input/Output as applied to the outer leads or interconnects of IC's or PWB's which interface with devices and assemblies in an electronic system.

Ion An electrically charged atom or group of atoms. Positively charged ions have a deficiency of electrons while negatively charged ions have a surplus of electrons.

Ion Exchange Resin Small particles which contain acidic or basic groups which will trade ions with salts in solutions.

Ionic Activity A measure of the ph. See ph.

Ion Implantation The doping of semiconductors by a very precise and reproducible method to achieve specific characteristics.

Ion Migration The movement of free ions within a material or across the boundary between two materials under the influence of an applied electric field.

Ionizable Material A material whose electrons are readily lost from atoms or molecules thereby reducing the electrical resistance of the material.

Ionization A process by which electrons lost from or transferred to neutral molecules or atoms to form positively or negatively charged particles.

Irradiance The radiated power per unit area incident upon a surface; measured in watts/cm^2.

Isolation The ability of an electrical component or circuit to withstand interferences.

Isopak A pin-grid array of Kovar pins sealed in glass-to-Kovar plate.

Isostatic Press A hydraulic press which applies equal pressure in all directions.

Isotactic When molecules are polymerized in parallel arrangements of radicals on one side of the carbon chain.

Izod Impact Strength A measure of the toughness of a material under impact as measured by the Izod impact test.

Izod Impact Test A test for measuring impact resistance of a plastic material. A notched sample bar of plastic is struck by a pendulum and the force required to break the sample is a measure of the impact strength.

Jacket A plastic, rubber, or synthetic covering over the insulation, core, or sheath of a cable.

Jackscrew A screw which is attached to one half of a two piece contact which is used to pull both halves together or to separate them.

J-Lead A package lead which has been shaped to resemble the letter J.

Joint The location at which two adherends are held together with a layer of adhesive.

Joint, Lap A joint made by placing one adherend partly over another and bonding together the overlapped portions.

Joint, Scarf A joint made by cutting away similar angular segments of two adherends and bonding the adherends with the cut areas fitted together.

Joint, Starved A joint that has an insufficient amount of adhesive to produce a satisfactory bond.

Josephson Junction The close proximity of two superconducting materials, but separated by a thin insulating barrier, to act as an electric switch.

Josephson Superconducting Device Two superconductive ceramic materials which act as a Josephson device at low temperatures.

Jumper A direct electrical connection between two points on a printed wiring board or substrate. Used to complete a circuit temporarily or to bypass a circuit.

Jumper Cable A short cable used to interconnect two printed wiring boards or other devices.

Junction (1) A point in a circuit where two or more wires are connected. (2) A contact between two dissimilar metals or materials. (3) A region of transition between p and n type semiconductive materials. (As in transistors and diodes).

K (1) The symbol for dielectric constant. (2) Abbreviation for kilo (1,000). (3) A 1K memory chip. However, it contains 1,024 bits because it is a binary device based on powers of 2.

Kerf A slit or channel cut in a resistor during laser trimming or by an abrasive jet.

Kevlar A DuPont trade name for an aramid which is a manufactured fiber in which the fiber-forming substance is a long-chain synthetic aromatic polyamide. It is a lightweight, high modulus fiber used in plastic reinforcement.

Key A short, hardened pin which slides into a corresponding mating hole to guide the two mating halves in the engagement and assembly of a connector.

Keying A mechanical arrangement of guide pins and sockets, plugs, bosses, slots and grooves in a connector housing, shell or insert which allows connectors of the same size and type to be properly aligned without the danger of making a wrong connection and damage to the mating pins and sockets.

Keyway A groove or slot which assures the correct orientation and subsequent mating of two connector halves.

K Factor The coefficient of thermal conductivity. The amount of heat that passes through a unit cube of material in a given time when the difference in temperature of the two faces is one degree Celsius.

Kiln A high temperature furnace used for firing ceramic materials.

Kirkendahl Voids Voids which are induced at the interface between two different metals with different interdiffusion coefficients.

Kovar A nickel-iron-cobalt alloy. Formerly a Westinghouse trade name. Similar alloys are used in hybrid packages.

Lacing Cord A flexible, flat cord available in several different materials which are either coated or impregnated. They are used for tying cable forms, hook-up wires, bundles of wires and wire harness assemblies.

Laminar Flow A directed flow of filtered air which is moved constantly across a clean work-area in a direction parallel to the surface of the station.

Laminar Wave Soldering A solder wave having little or no surface turbulence.

Laminate To unite sheets of material by a bonding material usually with pressure and heat (normally used with reference to flat sheets).

Laminate Void Absence of resin in any cross-sectional area which should normally contain resin.

Laminated Plastics Layers of resin impregnated reinforcements bonded together by heat and pressure to form a single sheet. Resins include phenolics, melamines, epoxies, silicones, polyesters, polyimides, and others.

Land A widened area of a conductor on a substrate which is used for the attachment of wire bonds or mounting and bonding of chip devices.

Land Pattern A group of lands which are used for chip mounting, wire bonds, or testing.

Lanyard A device used to uncouple connector halves by pulling a cord or wire. Only certain connectors are equipped with this device.

Lap Joint The joining of two conductors or other materials by placing them side by side so that they overlap.

Lapping A grinding and polishing process used to obtain a precise thickness or extremely flat, highly polished surface.

Large Scale Integration (LSI) An array of integrated circuits on a single substrate. Usually more than 100 gates or circuits and at least 500 elements.

Laser A device which produces a light beam by the emission of energy that is stored in a molecular or atomic system when stimulated by a signal. Light Amplification by Stimulated Emission of Radiation.

Laser Bonding A process in which a metal-to-metal bond is formed by the use of a laser beam as the heat source.

Laser Pattern Generation A method in which a precision laser beam is used as the light source to produce photo images for original artwork or direct pattern generation for integrated and other electrical circuits.

Laser Soldering A selective soldering process which uses a programmable laser system for soldering of wire wrapping to the backplanes, printed circuit boards, etc.

Laser Trimming Adjusting the resistor values by using a focused laser beam to cut and vaporize the resistor material. The resistor value increases as the resistor is trimmed.

Laser Welding A welding process in which the thermal energy is supplied by a laser impinged on the surface of the metal to be welded.

Latent Curing Agent A curing agent that produces long-time stability at room temperature but rapid cure at an elevated temperature.

Lattice Structure Of a crystal; the stable arrangement of atoms and their electron-pair bonds.

Layer A single film or level in a multiple film or level structure on a substrate.

Layout The positioning of conductors and resistors on artwork prior to photoreduction of a layout to obtain a negative or positive used in subsequent fabrication.

Lay-Up The process of registering and stacking layers of a multilayer board in preparation for the laminating cycle.

L-Cut A trim notch which is cut in a thick film resistor. It is made by the cut starting perpendicular to the resistor length and then turning 90 degrees to finish the trim parallel to the resistor axis.

Leaching The dissolving or alloying of a metal to be soldered into the molten solder.

Lead A wire that connects two points in a circuit and which is usually self supporting.

Leaded Pertaining to electronic devices which have electrical leads extending from their enclosures.

Leaded Chip Carrier (LDCC) A square or rectangular shaped ceramic form with a cavity to house a semiconductor device and with high density I/O pads and leads on all four sides of the package.

Leaded Chip Carrier Plastic (PLCC) A electronic package, made of plastic, containing a chip which has leads extending from the sides of the package in a typical J shaped configuration for surface mounting on printed circuit boards.

Leaded Surface Mount Component See Surface Mount Component (Leaded).

Lead Frame A rectangular metal frame with leads. The frame contains the leads which are connected to dies. After encapsulation of the DIP/package, the frame is cut off, leaving the leads extended from the package.

Leadless Pertaining to electronic devices which do not have electrical leads extending from their enclosures but rather solder lands or bumps located on top, bottom, or sides of the package.

Leadless Chip Carrier (LCC) A square or rectangular shaped ceramic form with a cavity to house a semiconductor device or other electronic components with metallized interconnecting lands. It is reflow soldered to a PWB or substrate.

Leadless Device A device having no input/output geometrical shaped leads.

Leadless Inverted Device (LID) A square or rectangular shaped ceramic carrier with a cavity used to house a semiconductor chip. It has metallized conductor lands for interconnections to the semiconductor device and metallized pedestals for reflow solder bonding to a PWB or substrate. It is normally coated with a plastic to protect the bonds and chip device. It is mounted inverted on the PWB or substrate.

Leadless Pad Grid Array (LPGA) A leadless SMT package containing I/O connections through metalized pads arrayed in a matrix on its base. Used primarily for high pin count IC's.

Lead Wires Wires used for intraconnections or input/output leads.

Leakage Current A small, stray, undesirable amount of current which flows through/across an insulator between two electrodes.

Leak Check A method for testing the hermeticity of a package. There are two types which are used for testing; gross leak and fine leak.

Leno Weave A leno weave is a locking type weave in which two or more warp threads cross over each other and interlace with one or more filling threads. It is used primarily to prevent shifting of fibers in open fabrics.

Leveling In thick film; the smoothing out or self leveling of screen mesh marks of pastes after a pattern or circuit has been printed.

Levels of Electronic Packaging and Interconnecting The various thresholds of connection points are:

First Level is between the devise and the package.

Second Level is between package to package.

Third Level is between PWB to PWB (motherboard or LRU).

Fourth Level is between the LRU and the system.

Lid A flat price of metal or ceramic used to seal packages. Stepped or depressed lids have a flange to conform to the inside of the package.

Life Aging A burn-in test conducted at an elevated temperature for extended periods of time to test the quality of the product.

Life Cycle A test which is indicative of the time span of a device before failure. It is conducted in a controlled, accelerated environment.

Life Drift A permanent change in the value of a device or level of a circuit element under load. Rated as a percentage change from the original value for 1,000 hours of life.

Life Test The test of a component under similar conditions at which the component operates, or simulates by acceleration to determine the life expectancy of the component.

Lift-Off Mark The image of the bond area which remains after the bond has been removed.

Light Emitting Diode (LED) A p-n or semiconductor device specifically designed to emit light when forward biased.

Lightning Electromagnetic Pulse (LEMP) A disturbance created by the electromagnetic field from lightning.

Limited-Coordination Specification (or Standard) A specification (or standard) which has not been fully coordinated and accepted by all the interested activities. Limited-coordination specifications and standards are issued to cover the need for requirements unique to one particular department. This applies primarily to military agency documents.

Linear A straight chain organic molecule, with linear carbon-to-carbon linkages. Also sometime called aliphatic. Some branching often exists off the linear chain.

Linear Circuit A circuit in which the output voltage is approximately directly proportional to the input voltage - usually a limited range of signal voltages and frequencies.

Linear Coefficient Of Expansion See Thermal Expansion.

Line Definition In thick film, the capability of producing sharp, clean screen printed lines. The precision of line width is determined by twice the line edge definition/line width. A typical precision of 4% exists when the line edge/definition width is 2%.

Line Discontinuity A point on a transmission line equivalent to a separate circuit having resistance, capacitance, and inductance, which produces false reflections.

Line Impedance The impedance as measured across the terminals of a transmission line.

Line Loading The connecting of external resistance, inductance, and capacitance in a transmission line.

Line Replaceable Unit (LRU) An electronic subassembly, equipped with a multi pin connector, which can readily be removed and replaced with another identical subassembly.

Line Resistance The resistance offered by conductor lines in a package. It is measured in ohms per unit length or for a given cross section in ohms per square.

Lines Conductor runs of an interconnection network.

Lines Per Channel The number of conductive lines between metallized holes or lands in a printed wiring board or a ceramic substrate.

Liquid Crystal Display (LCD) A visual display in which a liquid crystal material is hermetically sealed between glass plates. The application of a magnetic field, voltage, or heat alters the crystalline properties rendering segments visible by color contrast.

Liquid Injection Molding A fabrication process in which catalyzed resin is metered into closed molds.

Liquidus The temperature of a metal or alloy at which it is completely liquid.

Live-Line Connector A type of connector which may be installed or removed, using an insulated tool, with the power remaining on.

Loadbreak Connector A type of connector which is designed to open and close while current is flowing through the circuits.

Load Life That period of time over which a component can sustain its full power rating.

Locator Same as Stop Plate.

Locking Spring Same as Contact Retainer.

Logic Design The selection and interconnection of logic units to achieve logical function in the digital system.

Logic Part A group of logic circuits which are interconnected and packaged and used throughout a digital system.

Logic Primitive A basic logic function as it exists as a single unit.

Logic Service Terminal (LST) A terminal which carries logic signals.

Longitudinal Indent An indent whose longest dimension is in line with the connector barrel.

Loop The curvature of the wire between each end of the wire bonds.

Loop Height The maximum perpendicular distance from the top of the wire loop to a point between the two attachment wire bonds.

Loss Factor The rate at which heat is generated in a insulating material. It is equal to its dissipation times the dielectric constant.

Loss Tangent The amount of power lost as heat when a dielectric or semiconductor material is placed in an electric or electromagnetic field.

Lossy Signal Line A transmission line containing series resistance, skin effect, and dielectric conduction.

Low Energy Surface Materials which have surface free energies below 500 ergs/cm^2 and low melting points. Not easily wetted or bonded

Low Frequency In the radio frequency spectrum, the band from 30 to 300 KHz.

Low Level Noise Tolerance (NTL) The maximum noise level in the receiver when the input signal is in its "down" state.

Low Loss Dielectric An insulating material that has a low power loss over long lengths which makes it suitable for transmission lines. Usually has low dielectric constant and dissipation factor values.

Low Loss Substrate A substrate with high radio frequency resistance and low energy absorption. Usually has low dielectric constant and dissipation factor values.

Low Pressure Laminates Laminates which can be molded and cured at 400 PSI or below.

Lug A device, that can be soldered or crimped to the end of a wire, which has a eye or fork used for mechanical attachment.

Macerate To chop or shred fabric for use as a filler for a molding resin.

Maintainability The ability of a part, subsystem, or system to continue to operate or restored to a specified level of operation when the maintenance is performed in accordance with recommended procedures and resources.

Major Weave Direction The continuous-length direction of a roll of woven glass fabric.

Manhattan Effect A problem which occurs to small chips during reflow soldering. Because of surface tension and unbalanced forms of soldering wetting, one end of a chip will rise up and cause a solder open at one end of the chip. Synonymous with Drawbridge and Tombstone (preferred).

Manhattan Length The wire length between terminals of a net or connection measured in the X and Y directions on the wiring plane of a package.

Manufacture To make or produce by hand or by machinery on a small or large scale, a usable product or part.

Mask A thin sheet of metal, a glass photographic plate, or a photographic negative or positive which contains circuit patterns used for making thick film screens, thin film deposition, or semiconductors.

Mass Spectrometer An instrument capable of making rapid analysis of chemical compounds by means of gas ionization.

Mass Splice In fiber optics, it is a technique in which all of the fibers are joined simultaneously without handling each fiber individually.

Mass Termination A termination process in which terminals pierce flat cable insulation without stripping the insulation, and cold flow mate with the conductors to form a metal to metal joint.

Master Artwork Same as Master Layout.

Master Batch Principle A blending of resistor pastes used in thick film technology to a nominal value of ohms per square. The nominal value is then the master control number.

Master Drawing A drawing which shows the dimensional limits or grid locations to any and all parts of a printed circuit, size, type, location of holes, and other pertinent information necessary to fabricate the product.

Master Layout The original layout of a circuit.

Master Pattern A one-to-one scale pattern used to produce printed circuit boards within the accuracy of the master drawing. Also called Production Master.

Master Slice A wafer containing many groups or clusters of devices. The wafer is subsequently diced into individual devices.

Mat A randomly distributed felt of glass fibers used in reinforced plastics lay-up molding.

Matched Metal Molding Method of molding reinforced plastics between two close-fitting metal molds mounted in a hydraulic press.

Matched Seal A glass-to-metal seal in which Kovar and glass, both having the same CTE's, are used to form hermetically tight seals.

Matching (Ratio matching in thick films) An indication of how closely two resistors (or capacitors) in a network approximate each other's value. Significant in resistor-ladder networks to know that the resistors are within a given ratio. Matching is expressed as a percentage of nominal resistance.

Mate The joining of two connector halves.

Matte Finish A surface finish on a material which has a grainy appearance.

Matrix An orderly two dimensional array with circuit elements such as wires, diodes, relays, etc. arranged in rows and columns.

Matrix Tray See Waffle Pack.

Mealing A separation of a conformal coating from the base material of a printed wiring board. It usually appears in the form of spots or patches and is observable as blisters under the coating. Also known as Vessication.

Mean Time Between Failures (MTBF) A term used to express the reliability level. It is the average time (in hours) between failures of a device, circuit, or system on a continuous operating basis.

Mean Time To Failure (MTTF) The average of the lengths of time to failure for parts of the same type and operated as a group under the same conditions.

Measling Discrete white spots or crosses below the surface of the base laminate that reflect a separation of fibers in the glass cloth at the weave intersection. Similar to Crazing.

Mechanical Pertaining to machinery or tools.

Mechanical Properties Material properties associated with elastic and inelastic reactions to an applied force.

Medium Scale Integration (MSI) Integrated circuits which have less than 100 gates or basic circuits and at least 100 circuit elements.

Mega A prefix which means it is to be multiplied by one million (1,000,000).

Melamines Thermosetting resins made from melamine and formaldehyde and possessing excellent hardness, clarity, and electrical properties.

Memory A component which provides ready access to data or instructions previously recorded such as in a computer, control system, etc. so as to make them respond to a problem.

Memory Adder A device which is added to the basic cycle per instruction rate of the central processor which is not readily available in the CP. This is due to requirements for data or instructions.

Memory Chip A semiconductor device which stores information for later use.

Meniscugraph Test A solderability test which records surface tension of the solder bath during the test. The test specimen is connected to equipment through a strain gauge which records the surface tension of the solder bath.

Mesh Size The number of openings or squares per linear inch in a screen. A 100 mesh screen has 100 openings per linear inch.

Metal A material which has high electrical and thermal conductivity at room temperatures and exhibits malleability and ductility characteristics.

Metal Inclusion A metal particle imbedded in a nonmetal material.

Metallization A conductive film, single or multiple layers, deposited on the surface of another material by either vapor or chemical deposition to perform electrical and mechanical functions.

Metallized Ceramic (MC) A ceramic substrate which has been metallized with either screened thick film circuitry or evaporated thin film metallization.

Metal Mask A thin sheet of metal in which holes are etched to a desired pattern.

Metal Matrix Composites Any metal or metal alloy reinforced by adding organic (continuous) or nonorganic (discontinuous) fibers. Matrices include aluminum, copper, titanium alloys. Reinforcement types include graphite and silicon carbide.

Metal Oxide Semiconductor (MOS) A process in which an oxide or nitride is used as a dielectric layer between a metal and semiconductor.

Metal Oxide Semiconductor FET (MOSFET) A family of devices which uses field effect transistors. The current flow is through a channel of n and p type semiconductor material, controlled by the electric field around a gate.

Metal Thick Nitride Semiconductor A thick silicon nitride or silicon nitride-oxide layer which is used in place of an oxide.

Metal Thick Oxide Semiconductor (MTOS) A device which has a thicker oxide layer outside the active gate area to reduce parasitic problems.

Metal-To-Glass Seal See Glass-To-Metal Seal.

Micro A prefix which means it is to be divided by one million (1,000,000).

Microbond A bond using a small diameter wire (0.001 inches), usually gold, to a conductor or to a die.

Microcircuit A miniature circuit with a high equivalent-circuit-element density on a substrate that performs a electronic-circuit function.

Microcomponents Miniature components such as chip transistors, capacitors, and resistors.

Microcracks A small thin crack which can only be seen with a microscope at magnifications approaching 100 times (X).

Microelectronic Hybrid Package (MHP) A geometric shaped enclosure containing a ceramic substrate with thick or thin film circuitry, discrete components, and wire bond interconnections. The substrate is bonded to the base of the enclosure which is equipped with terminals to provide electrical access to the inside of the enclosure. A lid/cover is soldered or welded to the top of the enclosure to form a hermetic seal.

Microelectronics A part of electronic technology in which electronic systems are realized by the use of miniature electronic parts.

Microfinish A surface finish on a material, free of porosity, peaks, and valleys, uniform in thickness and having a microinch/micrometer finish.

Microminiaturization The technique of packaging microminiature circuits and miniature components through the use of screening, vapor deposition, diffusion, and photoetching.

Micron A unit of length equal to 10,000 A, 0.0001 cm, or approximately 0.000039 in.

Micropositioner A tool which is used to accurately position a substrate or a device for subsequent bonding or trimming.

Microprobe A sharp-pointed probe with a positioning handle which is used for making temporary electrical contact to a device or circuit for testing.

Microprocessor An integrated circuit that provides in a single chip functions equivalent to those in a central processing unit of a computer.

Microsectioning A series of steps which include cross-sectioning, encapsulation, polishing, and etching used in the preparation of a specimen for microscopic examination.

Microstrip A type of transmission line configuration which consists of a conductor over a parallel ground plane separated by a dielectric.

Microstructure (1) The structure of a properly prepared specimen as seen at high magnification. (2) A structure comprised of micro size particles which are bound together.

Microwave A term referring to wave lengths of 0.3mm to 1.0 m and a frequency range of 1,000 to 300,000 megahertz.

Microwave Integrated Circuit (MIC) A miniature microwave circuit using hybrid circuit technology to form the conductors and attach the chip devices and components.

Migration An undesirable movement of metal ions, especially silver, from one location to another in the presence of moisture and an electrical potential.

Mil A unit of length equal to 0.001 inches or 0.0254mm.

Military Pertaining to the armed forces of the United States; Army, Navy, Air Force, Marines.

Miniaturization A technique of packaging by reducing the size or weight of parts and arranging them for maximum utilization of space.

Minimum Annular Ring The minimum width of metal, at the narrowest point between the edge of the hole and the outer edge of the terminal ring.

Minor Defect An irregularity or imperfection which does not prevent the usability of a unit for its intended purpose.

Minor Weave Direction The width direction of a roll of woven glass fabric.

Misalignment Loss In fiber optics, it is the loss attributed to the lateral or angular misalignment of the optical junction centerline.

Mislocated Bond A wire bond which has part of the bond area off the bonding pad.

Mixed-Component Mounting Technology A technology which utilizes both Surface-Mounting and Through-Hole technologies within the same electronic package and interconnection levels.

Mock Leno Weave An open type weave that resembles a leno and is accomplished by a system of interlacings that draws a group of threads together and leaves a space between the next group. The warp threads do not actually cross each other as in a real leno and, therefore, no special attachments are required for the loom. This type of weave is generally used when a high thread count is required for strength and at the same time the fabric must remain porous.

Modifier A chemically inert ingredient added to a resin formulation that changes its properties.

Modular Connector A connector which is capable of having similar/identical sections added onto the original connector to provide additional capabilities.

Module A subassembly in a packaging plan containing components, built to a standard size and equipped with standard plug-in or solder terminations.

Modulus Of Elasticity The ratio of unidirectional stress to the corresponding strain (slope of the line) in the linear stress-strain region below the proportional limit. For materials with no linear range a secant line from the origin to a specified point on the stress-strain curve or a line tangent to the curve at a specified point may be used.

Moisture Resistance The ability of a material to resist absorbing moisture, either from the air or when immersed in water.

Moisture Stability The stability of a circuit to function electrically in a high humidity environment.

Mold A medium or tool designed to form desired shapes and sizes. To process a plastics material using a mold.

Molded A part which has been formed by pouring or forcing liquid plastic material into a cavity.

Molded Circuits Electrical circuits which have been printed and etched on a flexible release material and subsequently transferred and bonded to a contoured surface such as the inside of a plastic cover or chassis. Also known as a 3-D Circuit because the circuits are in three dimensions.

Molded Plug A connector which has an electrical cable molded to its end.

Molding Cycle One complete operation of the molding press to produce a part.

Mold Release A lubricant used to coat a mold cavity to prevent the molded piece from sticking to it, and thus to facilitate its removal from the mold. Also called "Release Agent".

Mold Shrinkage The difference in dimensions, expressed in inches per inch, between a molding and the mold cavity in which it was molded, both the mold and the molding being at room temperature when measured.

Molecular Weight The sum of the atomic masses of the elements forming the molecule.

Monolithic Ceramic Capacitor A multilayer ceramic capacitor.

Monolithic Device A device whose circuitry exists in one die or chip.

Monolithic Integrated Circuit (MIC) An integrated circuit that exists on a single chip and contains no discrete components.

Monomer A small molecule which is capable of reacting with similar or other molecules to form large chainlike molecules called polymers.

Mother Board A printed circuit board used for plug-in cards or daughter boards and subsequent interconnecting terminations between them.

Multichip A device which is capable of performing the functions of several chips. See Multichip Package.

Multichip Integrated Circuit An integrated circuit whose elements are formed on several semiconductor chips and individually attached to a substrate or package.

Multichip Microcircuit A microcircuit made up entirely of active and passive chips which are individually attached to the substrate and subsequently interconnected to form the circuit.

Multichip Microelectronic Assembly A package containing several discrete active electronic devices mounted on a substrate, containing thin/thick film circuitry. These devices and circuits are interconnected by thermocompression or ultrasonic bonds.

Multichip Module A package containing several chips on a ceramic or other type substrate. This term most commonly applies to the use of VLSI chips.

Multichip Package An electronic package containing more than one chip. One related term, multichip hybrid (or simply hybrid) defines a package containing a hybrid combination of attached and deposited microcomponents, including chips. A second related term, multichip module, defines a package containing more than one very large scale integrated (VLSI) chips.

Multifiber In fiber optics, it is a coherent bundle of fused fibers which function mechanically as a single glass fiber.

Multilayer A printed circuit board containing several layers of etched circuit patterns, accurately aligned, bonded together, and electrically interconnected by plated through holes.

Multilayer Board A product consisting of layers of electrical conductors separated from each other by insulating supports and fabricated into a solid mass. Interlayer connections are used to establish continuity between various conductor patterns.

Multilayer Ceramic (MLC) A ceramic substrate containing multilayers of thick film circuitry separated by dielectric layers and interconnected by vias.

Multilayer Ceramic Capacitor A ceramic capacitor consisting of several thin layers of ceramic. After the layers are electroded and assembled, the assembly is fired.

Multilayer Printed Circuits Electric circuits made on thin copper-clad laminates, stacked together with intermediate prepreg sheets and bonded together with heat and pressure. Subsequent drilling and electroplating through the layers result in a multilayer circuit board.

Multilayer Substrates A thick film process in which alternate layers of conductors and dielectric ceramic pastes are printed and fired. The conductive layers are interconnected by vias.

Multi-Mode In fiber optics, it is a single fiber which is capable of propagating several modes of a given wavelength.

Multi-Mode Fiber A fiber which can propagate several modes of a specific wavelength.

Multiple Circuit Layout A layout consisting of an array of the same circuits on a substrate.

Multiple Conductor Cable Two or more conductors or wires which are insulated from each other and also insulated from the outer covering.

Multiple Termination Module (MTM) A device, which when heated, gang solders terminations as well as insulates and environmentally seals the terminations of a flat conductor cable.

Multiplexer An electrical component which allows two or more coincident signals to be transmitted over a single channel.

Multiplexing The combining of two or more signals from several channels into a single channel for transmission.

Munsell Color System A color specification system used in photography, color printing, painting, etc. which is based on Munsell hue, value and chroma when viewed under specific circumstances.

Mutual Capacitance The capacitance between two conductors with all of the remaining conductors electrically connected and considered to be grounded.

Nailhead Bond See Ball Bond.

Nano A prefix which means it is to be divided by one billion (1,000,000,000), or 1×10^{-9}.

Nanosecond A term used to describe the rise time of a signal or time delay and is equal to 10^{-9} seconds or 0.000000001 seconds.

NC Contacts Abbreviation for normally closed contacts with the power "on" in the control circuit.

Near Infrared Reflectance Analysis (NIRA) The spectral region having a wavelength of approximately 1 - 3 micrometers in which infrared spectroscopy analysis is performed. Typical IR spectrophotometers use a light source with wavelength from approximately 3 - 15 micrometers.

Neck Break Bond A wire bond which fails above the ball bond of a thermocompression bond.

Necking Localized reduction of the cross-sectional area of a tensile specimen which may occur during loading.

Negative Image A reverse print of the circuit in which the clear or transparent areas correspond to the circuitry or conductive elements while the dark or opaque areas correspond to nonconductive base material.

Negative Temperature Coefficient A condition in which the length of a material or resistance of a device decreases when the temperature is increased.

NEMA Standards Property values adopted as standard by the National Electrical Manufacturers Association.

Nest Same as Anvil.

Net Those terminals which are interconnected to a common dc power source in a package.

Network A combination of electrical elements such as interconnected resistors and capacitors to form interrelated circuits.

Next Higher Assembly In a drawing system, it is used to describe an assembly of the next higher level in the breakdown of that system.

Nick A term used to describe a cut or notch in a conductor or insulation.

Nickel A metallic element. Atomic number of 28. A metal which is corrosion and temperature resistant and used for alloying purposes.

Noble Metal Paste Thick film paste composed of noble metals such as gold, and platinum, which are highly resistant to corrosion and oxidation.

NO Contacts Abbreviation for "normally open" contacts with power "off" in the control circuit.

Noise (1) An undesirable, high frequency disturbance in an electrical system which modifies or effects the performance of a desired signal.

Nominal Value The specified value as opposed to the actual value.

Nonconductive Epoxy An epoxy resin with or without a filler. A filler is often added to improve its thermal conductivity as well as improving its thixotropic properties. Usually called a Thermally Conductive Epoxy.

Non Contaminating Compound A material which is inert to surrounding materials so as to prevent leaching, contamination, or degradation of material under various environmental conditions.

Nonfunctional Land A land located on a layer which is not connected to any conductive pattern.

Non-Ionic Same as Non-Polar.

Nonlinear Device A device in which an increase in applied voltage does not produce a proportional increase in current.

Nonlinear Dielectric A ceramic material which has a nonlinear capacitance-to-voltage characteristic. A barium titanate ceramic capacitor is a nonlinear dielectric.

Nonpolar A substance that does not ionize in water or ionizes very little.

Nonpolar Solvent A solvent which is not electrically conductive and will not dissolve nonpolar compounds such as hydrocarbons and resins.

Nonwetting A surface condition in which molten solder will not adhere to it.

Notch Sensitivity The extent to which the sensitivity of a material to fracture is increased by the presence of a surface in homogeneity such as a notch, a sudden change in section, a crack, or a scratch. Low notch sensitivity is usually associated with ductile materials, and high notch sensitivity with brittle materials.

Nuclear Electromagnetic Pulse (NEMP) The electromagnetic field which is formed from a nuclear blast.

Nugget A small area of recrystallized material at a bond interface.

Numerical Aperture (NA) In fiber optics it is the characteristic of an optical conductor to the degree of openness of the input acceptance cone at the end of the fiber or its acceptance of impinging light.

Nylon The generic name for all synthetic polyamides. These are thermoplastic polymers with a wide range of properties.

Occluded Contaminant One which has been absorbed by a material.

O Crimp An "O" shaped insulated support crimp.

Off Bond See Mislocated Bond.

Off Contact In screen printing, the preset space or gap between the screen and substrate. Also called the Breakaway.

OFHC Abbreviation for oxygen-free-high conductivity copper having a 99.95% minimum copper content. Normally used in electronic applications.

Ohm The unit of electrical resistance. The resistance through which one ampere of current will flow when a voltage of one volt is applied.

Ohmic Contact An electrical connection between two materials across which the voltage drop is the same in either direction.

Ohms Per Square The resistance of a square area, measured between parallel sides of both thin and thick film resistive materials.

OLB Outer Lead Bonds or the wire bonds which are made at the next higher assembly level.

Olefin A family of unsaturated hydrocarbons with the formula C_nH_n named after corresponding paraffins by adding 'ylene' or 'ene' to the stem, i.e., ethylene. Paraffins are aliphatic hydrocarbons. (See Hydrocarbon.)

Olyphant Washer Test A simple and practical test developed by Murray Olyphant of 3M Company for testing the comparative crack resistance of casting and potting resins.

Onsertion The positioning of a surface mounted component onto the surface of a printed wiring board or substrate with respect to the solder lands. This is important and must be accurate.

Opacity The ability of a material to resist the entrance or penetration of light waves.

Open Barrel Terminal A terminal with an open insulated barrel to accommodate a wire and to be subsequently crimped to form a reliable electrical connection.

Open Cell Material A foamed or cellular material which is made up of cells and are interconnected. (Closed cell material are not interconnected.)

Open Entry Contact A female type of contact which is unprotected from damage caused by probes and other devices.

Operator Certification A program whereby operators are trained and certified to perform specific functions; (such as wire bonding, chip mounting, soldering, thick film screen printing, etc.).

Opposed Electrode Welding A resistant weld where the parts to be welded are placed between two opposed electrodes. The current is passed between the electrodes through the parts which are heated (and welded) due to the resistance of the parts interfaces.

Optical Comparator A microscopic device which is capable of projecting a magnified image of the work piece on a screen, and the dimensional measurements, surface flows, etc. can easily be made or measured.

Optical Conductors In fiber optics, materials which have a low optical attenuation to the transmission of light energy.

Optical Coupling In fiber optics, the leakage of light from one fiber into another. Often referred to as Crosstalk.

Optical Interconnects These are essentially optical-electronic type devices such as a (LED) light emitting diode and an optical coupler. The LED converts an electrical signal into light at the electro-optical interface while the optical coupler couples signals from one electronic circuit to another by the use of an LED plus an phototransistor. This combination converts an electrical signal of the primary circuit into light and the phototransistor in the secondary circuit, reconverts the light signal back into an electrical signal.

Optical Mosaic In fiber optics, a construction of fibers into a group or groups. However, some degree of imperfection occurs at the boundaries of the subgroup. When imperfection reach a high level it is called whichen wire.

Optoelectronic Defines a device which responds to optical power or uses optical radiation in its operation; a device which acts as an optical-to-electrical or electrical-to-optical transducer. Examples are photodiodes, LEDs, etc.

Orange Peel A rippled surface texture of solder mask material over circuits containing tin or tin-lead plating on PWB's. The orange peel indicates the reflow of the plated metals under the solder mask material.

Organic (1) Composed of matter originating in plant or animal life, or composed of chemicals of hydrocarbon origin, either natural or synthetic. Used in referring to chemical structures based on the carbon atom. (2) A chemical structure built upon the carbon atom. Most polymers are organic. Silicones, as an example of partially inorganic polymers, have a chemical structure which is built around the silicon atom. Organics burn to a black carbon ash, while silicones burn to a whitish silicon dioxide ash.

Organic Composites The family of fiber architectures that include continuous or discontinuous fibers in an organic resin matrix material, such as graphite-epoxy composites. Applications include commercial and military products that must be light weight, strong, and rigid.

Organic Pigments A dye or coloring material used to color plastics and having excellent resistance to color change.

Organic Vehicle The resin base material of a solder flux.

Organic Water Soluble Flux See Flux, Organic Acid (OA).

"O" Ring A circular shaped ring made of rubber which is used in very broad sealing applications.

Outer Lead Bonding The soldering of the peripheral leads of an electronic package, when using the tape automated bonding process, to the printed wiring board.

Outgas The release of gases and vapors from a material over a period of time when subjected to temperatures above approximately 100 degrees C or vacuum conditions or both.

Ovaled An elliptical shaped terminal or contact to accommodate two wires.

Oven Soldering A process in which multiple solder terminations are made simultaneously but with temperature limitations because few materials and components can withstand high temperatures.

Overbonding See Chopped Bond.

Overcoat A thin film of dielectric material which is applied over circuitry and components for environmental, mechanical, and contamination protection.

Overcurrent The excessive amount of current in a conductor which causes a rise in temperature of the conductor and its insulation.

Overflow Wire A wire connection which is called out on the wiring diagram but not actually connected during automatic wiring of the package.

Overglaze A printed and fired glass layer which is applied over resistors, or used as a solder barrier in thick film technology.

Overlap In thick film, it is the overlay or contact area between the film resistor and conductor.

Overlay The application of one material over another material.

Overpotential A voltage which is greater than the normal operating voltage of a component or circuit. Also called Overvoltage.

Overspray The undesired spreading of the abrasive from the nozzle of the resistor trimming machine onto adjacent resistors.

Overtravel (1) In screen printing it is the excess distance a squeegee travels in the "Y" direction, beyond the pattern on the substrate, before the squeegee lifts off the screen. (2) The excess distance in the "Z" or downward direction a squeegee blade would push the screen if the substrate and nest plate were not in place.

Overvoltage A voltage greater than the normal operating voltage. (Also called Overpotential.)

Oxidation (1) A process in which a metal reacts with oxygen in the atmosphere to form an oxide such as iron reacts with oxygen to form iron oxide or rust.

Oxidizing Atmosphere An air or oxygen enriched atmosphere in a furnace.

Ozone Test A test in which materials are exposed to high concentrations of ozone which produces accelerated indication of degradation.

Package An enclosure for electronic components and hybrid circuits consisting of a header, a lid, and hermetic sealed feed through terminal leads. Packages are made of metal, ceramic and plastic.

Package Crossing An electrical connection from a terminal on one package to a terminal on another package.

Package Delay The time delays caused by the interconnections and distance between components in order to complete their functions. This delay is dependent on both distance and materials used in the package.

Package Level Refers to the various members which make up the packaging hierarchy such as the chip, chip carrier, PWB, motherboard, chassis, system (in the order from a low to a high level).

Package Lid See Lid. Also called Package Cap.

Packaging The process of physically locating, connecting, and protecting electronic components.

Packaging And Interconnecting Assembly An assembly which has components mounted on one or both sides of an interconnecting structure (such as a printed wiring board).

Packaging And Interconnecting Structure Essentially a printed wiring board which consists of a composite board material with printed wiring.

Packaging Density The quantity of components, interconnections, and mechanical devices per unit volume. Classified as high, medium, or low densities.

Pad (1) A metallized area on a substrate to which wire bonds or test probes can be made. (2) A portion of the conductive pattern of printed circuits for mounting components. (Also called Land).

Pad-Grid Array Package A package with screened solder contact pads over the entire bottom in a checkerboard array. Each pad is located where a via exits the ceramic. Used in high I/O VHSIC devices.

Panel (1) A metal plate on which connectors are mounted. (2) The base material from which one or more printed wiring boards are made after being processed through printed wiring board processes.

Panel Mount The attaching of the female half of a connector to a panel.

Panel Plating The plating of the entire surface of a panel including holes.

Parallel Gap Solder A method in which excessive current is passed through a high resistance gap between two electrodes which reflows the solder and forms an electrical connection.

Parallel Gap Weld A weld which is formed by passing excessive current through a high resistance gap between two spring loaded electrodes, which in turn applies a mechanical force to the component leads or conductors.

Parallelism The amount of variation in thickness of a substrate, wafer, etc.

Parallel Splice A splice in which a holding tool is used to insure the conductors lie parallel to each other during joining.

Parasitic Losses Electrical losses in a circuit caused by the packaging and materials used.

Part An article that cannot be disassembled without destroying its intended use. (A transistor, screw, a substrate with conductors and resistors.)

Partial Discharge An electrical discharge which only partially bridges the insulation between conductors. Often referred to as Corona.

Partial Discharge Pulse A voltage or current pulse as a result of a partial discharge.

Partial Lift A wire bond partially raised from the bonded pad.

Parting Agent A lubricant which is applied to the surface of the mold to prevent the finished parts from adhering to the mold. Also called Release Agent and Mold Release.

Parting Line The line formed by the mating surfaces of the mold halves.

Partitioned Mold Cooling The cooling of the mold by circulating water through the core of the mold.

Parts Density The number of parts per unit of volume.

Parylene A polymer resin, polyparaxylene, which provides a very thin (250 - 500 angstroms), uniform conformal coating on printed wiring assemblies and electronic components. It is applied by vacuum deposition and provides a very uniform coating on sharp edges, complex shapes, and in holes.

Passivation (1) The application of an insulating layer of glass, SiO_2, or nitride over circuits and circuit elements for protection against moisture, contaminants, etc. (2) The growth of an oxide layer on the surface of a semiconductor to provide electrical stability by isolating the transistor surface from electrical and chemical conditions in the environment.

Passive Component An electrical component which does not change its character when an electrical impulse is applied. (Resistors and capacitors.)

Passive Network (1) A network which has no source of energy. (2) A network of passive elements (screened resistors) which are interconnected by conductors.

Passive Substrate A substrate which acts as a support of the circuitry or as a heat sink and does not have any active devices. Usually made of alumina, ceramic, or glass.

Passive Trimming Adjusting the function of a circuit while the power is turned off.

Paste A thick film screen printable composition of micron size polycrystalline solids suspended in a thixotropic vehicle. Often referred to as "ink" or "composition". They contain metals, metal oxides, and glasses.

Paste Blending See Blending.

Paste Soldering A process in which a paste, composed of solder particles in a flux paste, is screen printed onto a film circuit and reflowed to form connections to chip components.

Paste Transfer To pass a thick film ink or composition through a screen or mask and deposit it in a pattern on a substrate.

Path A part of a printed circuit between two pads or a pad and a terminal area.

Pattern The configuration of conductive and nonconductive materials on a substrate.

Pattern Plating The selective plating of a conductive pattern.

Peak Firing Temperature The maximum temperature in the firing profile of thick film pastes.

Peel Strength A measure of adhesion between two materials when performed in a pulling or peeling fashion and measuring the force required to separate the two materials. The units are measured in oz/mil of width or pounds/inch of width.

Penetration The entering of one part or material into another.

Percent Defective Allowable (PDA) The maximum percent of defective parts that will permit the lot to be accepted after the specified 100% test.

Percussive Arc Welding A process in which a fixed gap is maintained between the surfaces of the two parts to be welded while RF (radio frequency) energy is applied.

Perimeter Sealing Area The sealing surface on an electronic header. It is located along the perimeter of the header cavity and defines the area to which the lid or cover is bonded or welded.

Permanence The resistance of a given property to deteriorating influences.

Permeability The property of a material which allows the diffusion or passing of a vapor, liquid, or solid through the material without physically or chemically affecting it.

Permittivity Same as Dielectric Constant or Specific Inductive Capacity.

Ph A measure of the acidity or alkalinity of a substance, neutral being at a ph of 7. Acids range from a 0 to 7 while alkaline or base solutions range from a 7 to 14.

Phase Diagram A graphical representation of the compositions, temperatures, and pressures at which the heterogeneous equilibria of an alloy system occur.

Phenolic (1) A synthetic resin produced by the condensation of an aromatic alcohol with an aldehyde, particularly of phenol with formaldehyde. (2) One of the largest volume-produced thermosets. Extensively used because of their low cost and insulation characteristics.

Phenylene Oxide Based Resins A thermoplastic with a very low specific gravity. They are tough, rigid materials with excellent mechanical and dimensional stability up to 300 degrees Fahrenheit with low creep and moisture absorption. Electrical properties include a high dielectric strength and a low dissipation factor up to 1 MHz.

Phenysilane A thermosetting copolymer of silicone and phenolic resin; furnished in solution form.

Phosphor Bronze An alloy of copper, tin, and phosphorus.

Phosphosilicate Glass (PSG) A phosphorous doped silicon dioxide which is sometimes used as a dielectric layer because it prevents the diffusion of sodium impurities. Since it softens and flows at approximately 1050 degrees Celsius, it creates a smooth finish for subsequent layering.

Photo Etch A process in which circuit patterns are formed by exposing (polymerizing, light hardening) a photosensitive material (photoresist) through a photo positive or negative of the circuit and etching away the part of the film that was not protected by the polymerized material.

Photolithography A technology which is employed to create a pattern which includes rubylith, photoreduction, step and repeat, computer aided design and electron-beam techniques. This process will produce a mask that will be used to image a microelectronic device.

Photon The smallest unit of radiant energy.

Photopolymer A polymer that changes characteristics when exposed to light of a given frequency.

Photoresist (Negative) A light sensitive material which hardens or polymerizes when exposed to UV light and is resistant to chemical etching solutions.

Photoresist (Positive) A light sensitive material which breaks down and is easily removed in photographic developing solutions. The exposed area of metal is then etched away.

Physical Design The location and orientation of chips, packages, devices and the respective interconnections in a package or in a total system.

Pick and Place A manufacturing operation in which chips are correctly placed and oriented on their respective pads on the substrate utilizing an assembly aide, prior to bonding.

Pico A prefix which means it is to be divided by $\quad$ 1 million million (1,000,000,000,000) or 10^{-12}.

Pigment Finely divided powdered particles used for coloring and insoluble in the vehicle in which it is used.

Pigtail (1) In wire bonding, the excess wire which remains after a bond is made and extends beyond the bonding pad. (2) A short length of wire used as a jumper, ground, or for terminating purposes.

Pin A small diameter metal rod used as an electrical terminal and/or a mechanical support. They are used inside a package to support a wire bond and externally as a plug-in type connection. The are either straight or modified as a nail-head, upset-pierced, or a formed variety.

Pin Contact A male type contact designed to mate with a female contact.

Pin Density The number of pins on a printed wiring board per unit area.

Pin Grid A two-dimensional arrangement of electrically conductive pins equally spaced and parallel to each other.

Pin Grid Array (PGA) (1) A predetermined configuration of many plug-in electrical terminals for an electronic package or interconnection application. (2) A package with pins located over nearly all of its surface area.

Pinhole A small hole which extends through a printed element to the base material. It can be in both metallized and dielectric materials.

Pitch (1) The distance between a point on an image and the corresponding point on the corresponding image in an adjacent pattern lying in either a row or a column. Also known as center-to-center spacing. (2) The nominal distance from center-to-center of adjacent conductors.

Pits Small holes which do not extend through printed elements to the base material.

Placement The placement and correct orientation, whether manual or automatic, of IC's, packages, and cards in their respective locations at a given package level.

Plain Weave The plain weave is the most simple and commonly used. In this type of weave, the warp and filling threads cross alternately. Plain woven fabrics are generally the least pliable, but they are also the most stable. This stability permits the fabrics to be woven with a fair degree of porosity without too much sleaziness.

Planar Existing or lying in a single plane.

Planar Motor A motor having a flat planer configuration.

Planar Motor Voice Coil Servo A mechanical positioning device with very high accuracy, good feedback, limited excursion, and very high speed.

Plasma An electrically conductive gas composed of neutral particles, ionized particles, and free electrons which are used to dry etch or plasma etch dielectric materials.

Plasma Erosion Uncontrolled erosion of materials due to the high energy of ionized gas.

Plasma Etch A dry controlled etching process in which plastic substrates (epoxy glass PWB's) are exposed to ion bombardment by a gas in a vacuum to improve bondability during electroplating.

Plastic A polymer after being blended with all of the additives required for a final product. Additives may include plasticizers, flame retardants, fillers, colorants, etc.

Plastic Deformation A change in dimensions of an object under load that is not recovered when the load is removed; opposed to elastic deformation.

Plastic Device A package made of plastic, such as epoxies, phenolics, silicones, etc. and contains semiconductor or electronic components.

Plastic Encapsulation The embedding of an electronic assembly in a plastic material for environmental protection.

Plastic Range (1) A temperature range in which most metals can be worked or deformed without causing cracking. (2) A temperature range of a material in which the material is partially liquid and partially solid.

Plastic Shell A thin plastic container which is used to hold an electronic assembly during subsequent encapsulation. Also used as a container to provide environmental protection but not necessarily to encapsulate the assembly.

Plasticity A property of plastics which allows the material to be deformed continuously and permanently without rupture upon the application of a force that exceeds the yield value of the material.

Plasticize To soften a material and make it plastic or moldable, by means of either a plasticizer or the application of heat.

Plasticizer A material incorporated in a resin formulation to increase its flexibility, workability, or distensibility. The addition of a plasticizer may cause a reduction in melt viscosity, lower the temperature of second-order transition, or lower the elastic modulus of the solidified resin.

Plastisols Mixtures of vinyl resins and plasticizers which can be molded, cast, or converted to continuous films by the application of heat. If the mixtures contain volatile thinners, they are also known as organosols.

Plated Same as Electroplating.

Plated-Through Hole (PTH) A hole in which an electrical connection is made between internal and external conductive patterns, or both, by the deposition of metal on the wall of the hole.

Platens Flat mounting plates of a press, that can be adapted to supply heat or cooling, which apply a uniform temperature and pressure to the work piece.

Plating The process of chemically or electrochemically depositing metal on a surface. Copper, nickel, chromium, zinc, brass, cadmium, silver, tin, and gold are the most common electrodeposited metals.

Plating Anode See Anode.

Plating, Electroless A metal deposition process using a chemical reducing agent in a solution that is catalyzed by the metal. This process provides a very uniform thickness on irregular shapes and inaccessible cavities.

Plating, Electrolytic A metal deposition process in which an electrolyte (a solution containing dissolved salts of the metal to be plated) transfers cations from the anode into the electrolyte and onto the work piece or cathode by means of a direct electric current.

Plating Up A process in which the metal deposition is increased in thickness by electrolytic plating after the base material has been metallized with a thin conductive layer.

Plating Void A small hole or cavity in the plated surface caused by a metal inclusion of the base material or by contamination of the plating bath.

Platinum A precious, heavy metal which is white in color. Thermally stable at high temperatures and provides low, consistent surface resistance making it ideal for contact applications. It is resistant to corrosion and film formation and therefore can be used to replace gold plated metal parts.

Plug-In-Package An electronic package which can be plugged into or removed from a socket, printed wiring board, or other suitable connectors.

Point-To-Point Panel Wiring (1) A mechanized technique in which the wiring is run in a direct path from one terminal to another without dressing the wire. (2) A method of interconnecting components by routing wires between the connecting points.

Poisson's Ratio The absolute value of the ratio of transverse strain to axial strain resulting from a uniformly applied axial stress below the proportional limit of the material.

Polarity An electrical condition by which the direction of the current flow can be determined in a circuit.

Polarizing Slot A notch or slot which is machined into the edge of a printed wiring board, at an exact location, to assure accurate location and insertion of the mating connector.

Polar Solvent A solvent which can dissolve polar compounds such as inorganic salts. They contain hydroxyl or carbonyl groups and have a strong polarity. They cannot dissolve nonpolar compounds such as hydrocarbons and resins.

Polyamide Nylon.

Polyamide-Imide A plastic with outstanding thermal stability at high temperatures and good electrical properties. It has an aromatic structure which when heat cured forms a linear amide-imide homopolymer.

Polyarylsulfone A thermoplastic resin with good chemical, solvent, and impact resistance. It is composed of phenyl and biphenyl groups which are linked together by thermally stable ether and sulfone groupings.

Polybutadiene A thermosetting resin which can be transfer or compression molded. These resins have excellent electrical properties, good mechanical properties, and outstanding resistance to water and other liquids. They also are noted for their heat resistance, up to 500 degrees Fahrenheit, and temperature stability.

Polycarbonate Resin A thermoplastic resin with high strength and dimensional stability over a wide range of temperatures and humidity. Because of their outstanding heat stability, impact strength, and creep resistance these resins are excellent candidates for many applications.

Polycrystalline A material composed of a large number of crystals such as alumina, ceramic, and semiconductor materials.

Polyesters Thermosetting resins, produced by reacting unsaturated, generally linear, alkyd resins with a vinyl-type active monomer such as styrene, methyl styrene, or diallyl phthalate. Cure is effected through vinyl polymerization using peroxide catalysts and promoters, or heat, to accelerate the reaction. The resins are usually furnished in liquid form.

Polyethylene A thermoplastic material with excellent resistance to chemicals and moisture, flexibility at low temperature, high electrical resistivity, good dielectric properties at high frequencies and relatively low cost makes this material an excellent candidate for electrical insulation applications.

Polyimide A high-temperature thermoplastic resin made by reacting aromatic dianhydrides with aromatic diamines. It is used with glass fibers in the manufacture of printed circuit laminates and provides excellent resistance to wear and oxidation, high temperature stability, weathering resistance, and a low dielectric constant resulting in reducing propagation delay in multilayer construction.

Polyisobutylene A polymer of isobutylene with isoprene. Its resistance to oxidation, ozone, moisture, chemicals, and aging makes it a prime candidate for electrical apparatus and equipment application. Also called butyl rubber.

Polymer A compound formed by the reaction of simple molecules having functional groups which permit their combination to proceed to high molecular weights under suitable conditions. Polymers may be formed by polymerization (addition polymer) or polycondensation (condensation polymer). When two or more monomers are involved, the product is called a copolymer. Also, any high-molecular-weight organic compound whose structure consists of a repeating small unit. Polymers can be plastics, elastomers, liquids, or gums and are formed by chemical addition or condensation of monomers.

Polymerize To unite chemically two or more monomers or polymers of the same kind to form a molecule with higher molecular weight.

Polymer Thick Film (PTF) A thick film deposition formed by curing polymer based inks or pastes onto suitable substrates. Much lower curing temperatures (under 150 - 200 degrees Celsius) are involved than with cermet thick films, which require firing at 700 to 800 degrees Celsius.

Polymethyl Methacrylate A transparent thermoplastic composed of polymers of methyl methacrylate.

Polynary A material which has many ingredients. Binary systems have two compounds; ternary have three compounds, etc.

Polyphenylene Oxide A thermoplastic resin which has excellent electrical characteristics and dimensional stability from -275 to +375 degrees Fahrenheit.

Polyphenylene Sulfide Resin An aromatic polymer with excellent high temperature and chemical resistance. Service temperature to 450 degrees Fahrenheit and melting temperature of 550 degrees Fahrenheit. They can be filled with glass-reinforced fibers for electrical applications.

Polypropylene A plastic that is made by the polymerization of high purity propylene gas in the presence of organometallic catalyst at relative low pressures and temperatures. Noted for its high tensile strength, abrasion, moisture, and heat resistance.

Polystyrene A thermoplastic produced by the polymerization of styrene. It has excellent electrical properties, good dimensional stability, and moisture resistance.

Polysulfone A thermoplastic resin which is flame resistant, heat resistant (over 300 degrees Fahrenheit for extended periods of time), and dimensional stability at elevated temperatures.

Polyurethane Resins A family of resins used to form thermosetting materials by reacting them with water, glycols, or other urethanes.

Polyvinyl Acetate A thermoplastic material composed of polymers in vinyl acetate in the form of a colorless solid.

Polyvinyl Butyral A thermoplastic material which is derived from butyraldehyde. It is a tough, colorless, flexible solid, and used extensively in laminates, adhesives, coatings, and crosslinked with ureas and phenolics.

Polyvinyl Chloride (PVC) A thermoplastic material composed of polymers of vinyl chloride. It can be blended with other polymers to impart abrasion resistance, heat stability, low shrinkage, and moisture resistance. It can be converted into a colorless sheet or film by heat and pressure.

Polyvinyl Chloride Acetate A thermoplastic material composed of copolymers of vinyl chloride and vinyl acetate. It is a colorless solid material with good water, acid, and alkaline resistance.

Polyvinylidene Chloride A thermoplastic material composed of polymers of vinylidene chloride (1,1 - dichloroethylene).

Porcelain A glassy, vitreous, ceramic material. Many varieties are available such as alumina porcelain, boronsilicate, cordierite porcelain, forsterite porcelain, stealite porcelain, titania porcelain, and zircon porcelain.

Porcelain Enamel Technology The art of coating and bonding a vitreous, glassy, inorganic material to metal by fusion at temperature above 800 degrees Fahrenheit.

Porcelainized Steel Substrates Steel plates or substrates coated with porcelain. They are subsequently metallized to form dimensionally stable deposited or etched circuits.

Porosity Multiple voids in a material.

Positive Image The true or exact picture of a circuit pattern.

Positive Temperature Coefficient A condition of a material in which other properties such as the physical dimensions, resistance, etc. increase as the temperature of the material is increased.

Positive Temperature Coefficient Resistor A type of resistor whose resistance increases as the temperature of the resistor increases.

Post See Terminal.

Post Curing The additional curing at the cure temperature or elevated temperatures to fully cure and achieve maximum properties of a plastic.

Post Firing The refiring of thick film circuits to sometime change the values of resistors.

Post Stress Electrical To apply electrical power to a film circuit in order to stress the resistor and evaluate the change in values.

Post-Type Terminal A type of terminal which, after the wire has been wrapped around, a threaded nut or terminal type device is added to secure the wire to the terminal.

Pot To embed a component or assembly in a liquid resin, using a shell, can, or case which remains as an integral part of the product after the resin is cured. (See Embed and Cast.)

Potential Or voltage. The difference in voltage between two points in a circuit when one of the two points is at a zero potential.

Pot Life The time during which a liquid resin remains workable as a liquid after catalysts, curing agents, promoters, etc.,are added; roughly equivalent to gel time. Sometimes also called Working Life.

Potting Cup A form attached to the rear or back of a receptacle and provides a mold-like cavity for potting the wires and wire entry of the assembly.

Potting Mold An enclosure which is used to fill with the potting material and allowed to cure to seal the back of an electrical connector.

Power The time rate at which work is done; $P = W/t$ where W is work done in joules, t in seconds and P will be the power obtained in watts.

Power Cycling An accelerated reliability testing method in which a cyclic stress is applied to microelectronic components by applying electrical power, on an intermittent basis, to a specific heat generating component in that assembly.

Power Density In thick film thermal management, the amount of power dissipated from a film resistor through the substrate. It is measured in watts per square inch.

Power Dissipation The dispersion of heat generated in a device, component, circuit whose current flows through it.

Power Distribution Those conductors within a package which carry the electrical power to the circuits.

Power Factor The cosine of the angle between the voltage applied and the current resulting.

Power Hybrid Package (PHP) A geometrical shaped metal (copper) container with discrete components (transistors, resistors) which are soldered directly to the base of the container along with beryllia/alumina substrates which are bonded to the base. The container is equipped with large diameter hermetically sealed terminals to provide electrical access to the inside of the container. A lid/cover is soldered or welded to the top of the container to form a hermetic sealed package. It is designed to handle 40 - 80 watts of power per square inch compared to 4 -5 watts per square inch of power for a standard MHP.

Preform (1) A pill, tablet, or formed material used in thermoset molding. Material is measured by volume, and the bulk factor of powder is reduced by pressure to achieve efficiency and accuracy. (2) A geometrical shape, thin sheet of material such as solder, epoxy, etc., used for soldering or bonding. The bond/glue line thickness is more precisely controlled using preforms.

Preheating The heating of a compound prior to molding or casting in order to facilitate the operation, reduce cycle time, and improve the product.

Premix A molding compound prepared prior to and apart from the molding operations and containing all components required for molding: resin, reinforcement fillers, catalysts, release agents, and other compounds.

Prepolymers As used in polyurethane production, a reaction product of a polyol with excess of an isocyanate.

Prepreg Ready-to-mold sheet which may be cloth, mat, or paper impregnated with resin and stored for use. The resin is partially cured to a B-stage and supplied to the fabricator, who lays up the finished shape and completes the cure with heat and pressure. Also called "B" Stage.

Preseal, Visual A visual inspection of the entire hybrid circuit assembly, including components, wire bond interconnections, particulate, and any defects prior to sealing the package.

Press-Fit Contact A type of electrical contact that is forced into a hole of an insulator or conductor type base material.

Press-Fit Pin A connector pin which is forced into a hole of a substrate or printed wiring board to form a seal without the use of solders or a weld.

Pressure Contact An electrical contact which is made by a spring applied force.

Pressure-Bag Molding A process for molding reinforced plastics, in which a tailored flexible bag is placed over the contact lay-up on the mold, sealed, and clamped in place. Fluid pressure, usually compressed air, is placed against the bag, and the part is cured. Also called Bag Molding.

Pre-Tinned The application of solder to the leads of a component or to a wire prior to soldering the component, or wire in place.

Primary Side Of a printed wiring board or a packaging and interconnecting structure as indicated on the master drawing. This side contains the majority of the electrical components.

Primer A coating applied to a surface, prior to the application of an adhesive, to improve the performance of the bond.

Print And Fire In thick film terminology, a term used to describe the screening of the paste or ink on the substrate and subsequently firing in a high temperature furnace to a specific profile.

Printed Circuit A conductive pattern, which can be fabricated by several processes, bonded to the surface of a base material and used to interconnect electrical components to form a previously designed circuit.

Printed Circuit Board See Printed Wiring Board.

Printed Circuit Board Assembly A printed circuit board to which discrete components have been mounted. Also called a Printed Wiring Board Assembly.

Printed-Circuit Laminates Laminates, either fabric or paper-based, covered with a thin layer of copper foil and used in the photofabrication process to make lightweight circuits.

Printed Contact That portion of a printed circuit, usually near the edge of the board, which provides the contact area for a connector. Also called Terminal Area.

Printed Wiring A conductive pattern which is bonded to the surface or an inner layer of a base material which is used to provide a point to point connection only but does not include printed components.

Printed Wiring Board (PWB) A composite of organic and inorganic materials (resins and fibers) to form a base material onto which conductive patterns are bonded to the surface for interconnecting electrical components.. It can be single or double sided or a multilayer construction of either rigid or flexible composite materials. Also called Printed Circuit Board.

Printed-Wiring Substrate Conductive patterns which have been printed as a substrate, usually made of alumina, ferrite or other mechanical supporting materials.

Printing The process of reproducing an image or pattern on the surface of a material by photoetching, vapor deposition, screenprinting, or diffusion.

Printing Parameters Factors which can be controlled in the screening process of thick films such as breakaway, speed and pressure on the squeegee, etc.

Print Laydown In thick film technology, the screening of a circuit pattern on a substrate.

Probe A rigid, pointed, metal, wire shaped device used for making electrical contact to a circuit pad for electrical test purposes.

Processor (1) In hardware, a unit which processes data. (2) In software, a computer program which compiles, assembles, and translates related functions for a specific language including logic, memory and control.

Procuring Activity The agency (government, contractor, subcontractor) which contracts for the purchase of equipment, spare parts, services, etc., and has the authority to grant waivers, deviations or exceptions to the procurement documents.

Profile A graphic representation of time vs temperature of a continuous thick film furnace cycle.

Programmed Wiring A process in which wires are attached to a termination panel containing many posts by using programmable equipment.

Promoter A chemical, itself a weak catalyst, that greatly increases the activity of a given catalyst. Also called an Accelerator.

Propagation The movement or travel of electromagnetic waves through a medium.

Propagation Delay The delay in time between the input and output of a signal. The delay time is measured in nanoseconds per foot of conductor.

Propagation Velocity The velocity that a signal travels along a transmission line. It is measured as a percentage of the speed of light.

Property Any physical, chemical, or electrical characteristic of a material.

Proportional Limit The greatest stress a material can sustain without deviating from the linear proportionality of stress of strain (Hooke's Law).

Proton A positively charged particle equal to the negative charge of the electron but with a mass of 1846 times that of the electron.

Prototype A handmade working model which is representative of the final design that is used for evaluation and test.

Pull Strength See Bond Strength.

Pull Test A term often used to measure the bond strength of a lead, wire, or conductor.

Pulse A sudden change in current or voltage from one value to a higher or lower value and back to its original value in a specific time.

Pulse Vacuum It is a technique in which heat is applied to a solder joint by means of a soldering iron until the solder melts and a vacuum is then applied by means of a plunger/bulb and the molten solder is drawn away from the joint. Also known as Solder Sucker.

Pultrusion Reversed "extrusion" of resin-impregnated roving in the manufacture of rods, tubes, and structural shapes of a permanent cross section. The roving, after passing through the resin dip tank, is drawn through a die to form the desired cross section and subsequently through a heater to cure the composite.

Puncture Strength The voltage at which the dielectric or base material is punctured by the voltage stress at the point of puncture. Usually, the voltage level at puncture is not known, and steps must be taken to eliminate the feature which is causing the stress.

Purge To evacuate a chamber, containing components, packages, or devices, of moisture and other contaminates prior to sealing or backfilling with an inert gas.

Purple Plague A purple colored gold-aluminum compound which is formed in bonding gold to aluminum. The compound is activated by moisture, and temperatures exceeding 350 degrees Celsius and the presence of silicon. It causes serious degradation of semiconductor devices.

Push-Off Strength The force required to break the bond between a chip and its mounting pad by applying a force to one side of the chip and parallel to the mounting pad. The bonding area should be free of excessive filets in order to get an accurate reading.

Pyrolyzed A material which has undergone chemical decomposition by heat usually without oxidation.

Quad Flat Pac (QFP) A chip carrier made of ceramic or plastic and having its leads bent downward and away from the four sides of the package.

Quad In Line Package (QUIP) A plastic package similar to a dual-in-line package with leads extending out of the package on 1.27 mm centers. However, half of the leads are bent at the edge of the package while the remaining project pass the edges of the package 1.27 mm before being bent down, creating a staggering condition. The package can also have staggered leads to halve the effective center-to-center spacing and can be Surface or Through Hole Mounted.

Quadpack Same are Quad Flat Pac. An IC or hybrid IC with leaded or leadless connections on all four sides of the package.

Qualified Products List (QPL) A list of commercial products and materials which have been pretested and found to meet the requirements of a specification, especially government specifications.

Quality The conformance of a component system to a specification.

Rack And Panel Connectors A connector which connects the inside back end of a cabinet with the drawer containing the electrical equipment when it is fully inserted.

Radial Spread Coating A coating process in which a predetermined amount of resin is dispensed on the top surface of a chip or circuit board. After spreading over the entire surface, it is cured to form a solid protective coating. Also called Glob Top.

Radiation The emission, transfer, or propagation of energy in the form of electromagnetic waves or particles through space. Radiation can be thermal, radio, visual, and x-ray energy.

Radiation Hardened (1) A process by which components and circuits are pre-exposed to high gamma and neutron radiation so that their performance is not degraded when subsequently exposed to hostile environments. (2) Components and circuits that are designed and manufactured such that when exposed to gamma and neutron radiation, their performance is not degraded beyond useable limits. (Same as radiation resistant.)

Radiation Transfer Index (RTI) In fiber optics, the transmission performance of an optical fiber cable, including the coupling and propagation losses.

Radio Frequency Interferences (RFI) Electrical signals from internal or external sources which interfere with the operation of an electrical system or electronic equipment. Radio frequencies range from 30 KHz to 300,000 MHz.

Radiograph An x-ray photographic image of an object. An example would be a sealed electronic package, or a metallized hole in a ceramic substrate.

Random-Access-Memory (RAM) A memory in which information can be independently stored or retrieved. Usually bits of information are only stored temporarily.

Random Failure A failure which occurs on a random or unpredictable basis but the failure rate for the sample population being nearly constant. Also called Random Network.

Rapid Impingement Speed Plating (RISP) A proprietary plating process in which speed plating is accomplished by using high current density and forced movement of the plating solution. Both pushing and pulling of the solution is utilized.

Rated Temperature The maximum temperature an electrical component can operate for an extended period of time without changing its properties.

Rated Voltage The maximum voltage an electrical component can operate for extended periods of time without degradation.

RC Network A network composed entirely of resistors and capacitors.

Reactance (X) The opposition offered to the flow of alternating current by the inductance or capacitance of a component or circuit.

Reaction Etching A chemical process in which metallic circuits are formed by removing uncoated areas of a conductive pattern.

Reaction Injection Molding (RIM) A molding process in which proportional amounts of two materials are fed into a reaction chamber, thoroughly mixed together, and subsequently forced into a mold for polymerization.

Reactive Diluent As used in epoxy formulations - a compound containing one or more epoxy groups which functions mainly to reduce the viscosity of the mixture.

Reactive Metal Any metal which readily forms compounds.

Read Only Memory (ROM) A memory device in which information is permanently stored during its manufacture or installation and cannot be erased.

Rebonding-Over Bond To apply a second bond on the same pad area after the initial damaged bond has been removed or to place a second bond adjacent to the initial bond.

Receiver A system which converts electrical waves into audio or visible form.

Receiving Element In fiber optics, the accepting side of the termination at the conductor interface.

Receptacle Connector A type of connector which is designed for easy mounting on a panel, bulkhead, or chassis and mates with a plug connector.

Reducing Atmosphere An atmosphere to which a reducing gas has been added to prevent oxidation of metal parts while they are being fired.

Reduction Dimension An exact dimension which is placed on an actual size layout drawing, located between two marks, which is used to verify the exact distance after photographic reduction has been made.

Redundancy A design which employs additional components or circuits, more then necessary, to perform the function, in order to improve the reliability.

Reference Edge An edge from which measurements are made or a conductor can be identified with the number one conductor being the closest to the edge.

Refiring In thick film technology, it is the recycling of a film resistor through the firing cycle to change the resistor value.

Reflow The application of heat to a surface containing a thin deposition of a low melting metal or alloy, resulting in the melting of the deposition, followed by solidification. A common example is the reflow of deposited solder metal or solder paste.

Reflow Soldering A method in which solder metal or paste is first applied to the surfaces of the parts to be joined and subsequently heated, causing the solder to melt and reflow to form a solder joint.

Refractive Index The ratio of the velocity of light in a vacuum to its velocity in a substance. (2) The ratio of the sine of the angle of incidence to the sine of the angle of refraction.

Refractory Metal Metals such as molybdenum or tungsten, which have extremely high melting points. (Molybdenum has a mp of 2620 degrees Celsius.)

Registration The alignment of a circuit pattern on a substrate or the pattern with respect to other layers of a double sided or multilayer board.

Registration Marks The marks which are used for aligning successive layers of printed wiring patterns on multilayer boards.

Reinforced Molding Compound A plastic to which fibrous materials such as glass, cotton, or others has been added to improve certain physical properties such as flexural strength.

Reinforced Plastic A plastic with strength properties greatly superior to those of the base resin, resulting from the presence of reinforcing materials in the composition.

Reinforced Thermoplastics Reinforced molding compounds in which the plastic is thermoplastic.

Relative Humidity The ratio of the quantity of water vapor present in the air to the quantity which would saturate it at any given temperature.

Relay An electrical/mechanical device. It is electrically controlled to mechanically open and close electrical contacts in the same or another circuit.

Release Agent See Mold Release.

Reliability (1) The probability that a system or component will have a failure-free performance under certain environmental conditions for a period of time. (2) The continued conformance of a device or system to a specification over an extended period of time.

Rents Rule A relationship which states that the number of input/output terminations in a logic package is proportional to a fractional power of the number of logic gates interconnected in the package. The functional power can vary with circuit complexity.

Repair A manufacturing operation which restores a part or an assembly to an operable condition but does not eliminate nonconformance.

Resin A high-molecular-weight organic material with no sharp melting point. For general purposes, the terms resin, polymer, and plastic are often used interchangeably. (See Polymer.)

Resin Bath In the manufacture of fiber glass reinforced plastics, a resin filled container, in which the reinforcing materials are immersed and wetted with resin.

Resist A material such as an ink or paint which is used to protect selected areas during chemical etching, plating, and soldering. Examples are: solder resist, plating resist, and photoresist.

Resistance The property of an electrical conductor which determines the amount of current produced by a given difference of potential. The ohm is the unit of resistance.

Resistance Soldering A process in which an electrical current is passed through an electrode and creates heat. The heated electrode is then placed in contact with the solder which melts.

Resistance Welding The joining of two electrically conductive materials by heat and pressure. The heat is generated by passing electrical current through the two conductors which are held together mechanically until the weld is complete.

Resistivity The characteristic of a material that resists passage of electric current either through its bulk or on its surface. The unit of volume resistivity is the ohm-centimeter, and the unit of surface resistivity is ohm/square.

Resistor An electrical component made of material which has a known resistance and opposes the flow of electrical current.

Resistor Drift The change in resistor value of a resistor which takes place through aging (time). It is rated as percent change per 1,000 hours.

Resistor Geometry In thick films, it is the physical outline of a screened resistor.

Resistor Overlap In screen printing, it is the overlap contact area the film resistor makes with a film conductor.

Resistor Termination Same as Resistor Overlap.

Resistor-Transistor Logic (RTL) A logic circuit composed of several resistors, a transistor, and a diode.

Resolution See Line Definition.

Rework A manufacturing operation which not only restores a part or an assembly to an operable condition but to the requirements of the contract, specifications, drawings, etc.

R.F. Connector A type of connector to which a coaxial cable can readily be attached. It exhibits the same characteristics as the cable.

Rheology The study of the flow properties of viscous materials.

Rib A structural reinforcing member of a molded part.

Ribbon Cable A flat cable whose conductors are insulated from each other.

Ribbon Interconnect An electrical interconnection between circuits or to the output pins of a package in which a flat narrow ribbon conductors are used.

Ribbon Wire A flat, flexible, metal wire having a rectangular cross section.

Rigid Coating A conformal coating, usually a thermosetting resin, which does not contain any plasticizers to keep the coating pliable.

Rigidsol Plastisol having a high elastic modulus, usually produced with a crosslinking plasticizer.

Risers See Vias.

Rise Time The time required for the initial edge of a pulse to rise from 10% to 90% of its peak value.

Robotic A general term that describes a science of mechanical devices which are designed to perform a variety of tasks in order to replace a human being.

Rockwell Hardness Number A number derived from the net increase in depth of impression as the load on

a penetrator is increased from a fixed minimum load to a higher load and then returned to minimum load. Penetrators include steel balls of several specified diameters and a diamond-cone penetrator.

Roentgen The amount of radiation that will produce one electrostatic unit of ions per cubic centimeter volume.

Rosin Activated Flux (RA) See Flux, Rosin Activated (RA).

Rosin Flux (R) See Flux, Rosin (R).

Rosin Mildly Activated (RMA) See Flux, Rosin Mildly Activated (RMA).

Rosin Solder Connection A defective solder joint in which the connection is held together by an invisible film of flux. Also called a Rosin Joint.

Rotational Casting (or Molding) A method used to make hollow articles from thermoplastic materials. Material is charged into a hollow mold capable of being rotated in one or two planes. The hot mold fuses the material into a gel after the rotation has caused it to cover all surfaces. The mold is then chilled and the product stripped out.

Routing Program An automated interconnecting layout program.

Roving The term roving is used to designate a collection of bundles of continuous filaments either as untwisted strands or as twisted yarns. Rovings may be lightly twisted, but for filament winding they are generally wound as bands or tapes with as little twist as possible.

Rubber An elastomer capable of rapid elastic recovery; usually natural rubber, Hevea. (See Elastomer.)

Rubylith A laminate material consisting of a thin red film with a heavier clear backing. It is used to make master artwork by cutting and peeling away portions of the red layer.

Sapphire The monocrystalline form of alumina, Al_2O_3. An insulating material on which silicon can be grown and etched away to form a solid state device. Also known as Corundum and Alumina.

Scaling The separation of film conductors and resistors from the substrate.

Scallop Marks Screen printed lines having irregular/jagged edges which can be caused by incorrect squeegee pressure, insufficient emulsion thickness, or incorrect screen mesh size.

Scavenging See Leaching.

Schematic Diagram A drawing which shows only the graphic symbols, electrical connections, and components which makes up a circuit.

Scored Substrate A substrate which has been diamond scribed to form a thin cut line for subsequent breakage.

Screen A metal or fabric network of various size squares (mesh size) and mounted snugly on a frame. An emulsion is then bonded to the screen and circuit patterns and configurations superimposed by photographic means.

Screen Frame A rectangular configuration made of wood, metal, or plastic on which a screen is snugly mounted and held firmly in place.

Screen Printing A thick film process in which a paste/ink is squeezed through open areas of a screen and transferred to the surface of a substrate to form film circuits and configurations. Also called Screen Deposition.

Scribe To scratch with a hard, pointed material such as a diamond.

Scribe And Break A technique in which a hard pointed material, for example a diamond, is used to scratch a ceramic substrate and subsequently put under tension to break along the scratched line.

Scrubbing To apply circular action to a clean chip or substrate during the bonding operation to improve the wettability in the formation of the bond.

Sealing In hybrid operations, a process in which

the lid/cover is joined to the header to form a sealed bond.

Sealing Plug A plug which is placed in a contact opening of a connector insert, which is not used, to provide an environmental seal.

Search Height The height of the bonding tool above the bonding area at which final adjustments in the location of the bonding area under the tool are made prior to lowering the tool for bonding.

Secondary Side It is the side of a printed wiring board or a packaging and interconnecting structure opposite the primary side and is the side on which the soldering is performed when through-hole mounting technology is employed.

Second Bond The second bond of a bond pair made to form a conductive connection.

Second Radius The radius of the back edge of the bonding tool foot.

Selective Etching A process in which the etching is restricted by using a chemical which will attack only one of the exposed metals.

Selective Plating The electrochemical deposition of a metal on specific areas of a part. The other areas are covered with a masking or resist type material prior to plating.

Self Extinguishing A characteristic of certain materials to extinguish its own flame after the source of the flame has been removed.

Self-Heating The generation of heat caused by a chemical or exothermic reaction.

Self-Passivating Glaze A glossy finish which appears on the surface of thick film resistors after firing and seals the surface from moisture absorption.

Self Stretching Soldering Technology (SST) A technique used to increase the height of a solder joint. By using two different solder alloys having different thermal expansion characteristics or different size solder bumps, the expansion or surface tension of the non functional bumps is increased thereby stretching the height of the functional solder joint. This technique produces higher solder joints which can withstand greater thermal stresses.

Semiconductor A material whose electrical conductivity and resistivity is between a conductor and an insulator. (Materials such as germanium, lead sulfide, silicon, gallium arsenide, and silicon carbide.)

Semiconductor Carrier A protective structure which is used for mounting semiconductor devices and has metallized internal pads and external feed throughs for connecting the chip to a substrate or PWB.

Semiconductor Chip A square or rectangular piece of semiconductor material, which has been processed to form an electrical device such as an integrated circuit or transistor.

Semiconductor Device An electronic device made of semiconductor material, such as a transistor, diode, or an integrated circuit. Also often called a Chip.

Sequentially Laminated Multilayer Printed Wiring Board A multilayer board composed of double sided boards and/or multilayer boards which are bonded together and each containing plated through hole interconnections prior to bonding and having blind and/or buried vias.

Serpentine Cut A zigzag cut in a film resistor to increase the resistor length thereby increasing the resistance in the trimming process.

Serrations Miniature grooves on the inside diameter surfaces of a terminal wire barrel. These provide additional electrical conductivity and improved tensile strength after crimping.

Service Rating The maximum current and voltage that an electrical component is designed for and is capable of carrying on a continuous basis.

Set (Mechanical) Strain remaining after complete release of the load producing the deformation.

Set (Polymerization) To convert an adhesive or other resin forms into a fixed or hardened state by chemical or physical action, such as condensation, polymerization, oxidation, vulcanization, gelation, hydration, or evaporation of volatile constituents.

Shadow Effect During wave soldering as the assembly moves over the surface of the molten solder, the creation of a void, depression, or skip behind a large component so that solder does not reach the solder land.

Shape Factor For an elastomeric slab loaded in compression, the ratio of the loaded area to the force-free area.

Shear Rate The relative rate of flow of viscous fluids.

Shear Strength The maximum shear stress a material is capable of sustaining. In testing, the shear stress is caused by a shear or torsion load and is based on the original specimen dimensions.

Sheet A rolled flat material, insulating or conducting, usually less than 3/16 inches in thickness.

Sheet Molding Compound Compression-molding material consisting of glass fibers longer than 1/2 in. and thickened polyester resin. Possessing excellent flow, it results in parts with good surfaces.

Sheet Resistivity The material resistance of a thick film ink or paste, expressed in terms of ohms per square (ohms/square). The resistance of a material is given in ohms, according to the formula:

$$R = p(L/WT)$$

where p is the material volume resistivity (a constant for a given material), L is the material length, W, the width, and T, the thickness. For a rectangular layer of material whose length and width are the same, the thick-film material resistance, or sheet resistivity, becomes p/T, expressed as s. For a material film layer (sheet) of constant thickness, sheet resistivity is fixed, regardless of the size of the square. The actual resistance value depends on the sheet resistivity and length and width dimensions:

$$R = (s) \ (L/W).$$

Length divided by width is a dimensionless number known as the aspect ratio, and is given an arbitrary unit called a square.

Shelf Life An expression to describe the time a molding compound or other resin forms can be stored without losing any of its original physical or molding properties.

Shell The outermost case or enclosure which houses the insert and contacts of a connector and provides correct alignment and protection of projecting contacts.

Shielding A metallic covering used to prevent electromagnetic or radio frequency interference during receiving or transmitting of a signal.

Shielding Effectiveness (SE) The shielding effectiveness of a metallic barrier in a measurement of the reduction in field strength between a source and a receptor of electromagnetic or radio frequency energy.

Shore Hardness A procedure for determining the indentation hardness of a material by means of a durometer. Shore designation is given to tests made with a specified durometer instrument.

Shorting Plug Same as Dummy Connector Plug.

Shrinkage The decrease in the physical dimensions of a molded part through cooling or of a casting upon polymerizing.

Shroud Same as Insulation Support.

Shunt An electrical device, made of low resistance material, which is used to deliberately bypass part of a circuit.

Signal An electrical impulse of a predetermined voltage, current, polarity, and pulse width.

Signal Distribution The conductors within a package which interconnects the drivers and receivers.

Signal Wiring The conductors which carry the electric signals.

Silicones Resinous materials derived from organosiloxane polymers, furnished in different molecular weights. Includes liquid and solid resin forms.

Silicon Monoxide A material used in thin film technology as an insulating material and is vapor deposited on selected areas of thin film circuitry.

Silver (Ag) A metal with good electrical conductivity and corrosion resistance. It is relatively inexpensive and is used to electroplate copper conductors and is readily solderable.

Single Chip Carrier An enclosure which houses one chip and connects the chip terminals to the next higher level.

Single Chip Module (SCM) An enclosure or package which accommodates only one chip.

Single In Line (SIP) A package resembling a dual-in-line package except it contains only one line of leads instead of a double line of leads.

Single Layer Metallized Package (SLAM) A leadless package, which does not have a cavity, made of ceramic and sealed to a ceramic cap with a ceramic or glass bond.

Single-Sided Assembly A packaging and interconnecting assembly (PWB) with components mounted on the primary side only.

Sinking The electrical shorting of one conductor to another in the screen printing of multilayer circuits. It is caused by downward movement of the top conductor through the molten glass at the crossovers.

Sink Mark A depression or dimple on the surface of an injection-molded part due to collapsing of the surface following local internal shrinkage after the gate seals. May also be an incipient short shot.

Sintering A process of bonding metal powders together using pressure and subsequently firing them into a strong cohesive mass.

Skew Ray In fiber optics, a ray which does not intersect the axis of a fiber but rather travels around the fiber along the length and outside surface of the fiber.

Skin Effect The tendency of current to flow on the surface of a conductor. The resistance increases with the surrounding dielectric media.

Slump The spreading of a screen printed thick film paste after printing and prior to drying resulting in a loss of line definition. Generally caused by a low viscosity of the paste.

Slurry A mixture of solids, in suspension, in a liquid.

Slush Molding Method for casting thermoplastics, in which the resin in liquid form is poured into a hot mold where a viscous skin forms. The excess slush is drained off, the mold is cooled, and the molding is stripped out.

Small Outline Package (SOP) A small, rectangular, IC surface mounted device with leads on 1.27 mm, 1.0 mm, and 0.85 mm spacing.

Small Scale Integration (SSI) Integrated circuits having less than ten logic gates.

Smeared Bond A wire bond that has been enlarged because of movement of the bonding tool or the holding fixture.

Snapback In screen printing, it is the return of the screen to its normal plane after being deflected by the squeegee after it has moved across the screen and substrate.

Snap-Off Distance See Breakaway.

Snapstrate A substrate that has been laser or diamond scribed that can be broken apart easily and without damage.

Soak Time The length of time during the peak temperature of the firing cycle in which a thick film paste is held.

Softening Point For glass having a density of 2.5, this temperature corresponds to the log viscosity of 7.6 poises per ASTM Method C 338.

Soft Error A memory state error which is caused by a process but with no permanent change to the physical condition of the device.

Soft Glass A type of glass which has a low softening point, approximately 450 degrees Celsius. Contains a high percentage of lead and sometimes called "solder glass" because of its wettability to metal surfaces.

Soft Solder A solder alloy with a low melting point, generally below 800 degrees Fahrenheit (425 degrees Celsius).

Software The programs and instructions for a computer or microprocessor system.

Solder An alloy with a relatively low melting point which is used to bond two metals together, each having higher melting points than the solder.

Solderability The property of a metal to be easily wetted by solder and to form a strong bond with the solder.

Solder Acceptance Same as Solderability or Wettability.

Solder Balls Small spheres of solder which remain on the surface of a printed wiring board assembly after wave or reflow soldering. These must be removed because they are sources of electrical shorts.

Solder Bridge An electrical short circuit between two conductors by solder which bridges the dielectric and electrically joins the two conductors.

Solder Bumps Solder balls which are bonded to contact areas or pads of components which are subsequently used for face-down bonding.

Solder Contact A contact or terminal which has a cup into which a wire is inserted and soldered.

Solder Contact Terminal The location on a motherboard to which a connector is soldered.

Solder Cream Same as Solder Paste.

Solder Cup The end portion of a terminal into which the conductor is inserted and subsequently soldered in place.

Solder Dam A dielectric screen printable paste used for controlling molten solder from spreading on conductors.

Soldered Joint A metallic bond between a solder alloy and the surface to be bonded such as leads,pads, terminals, etc.

Solder-Eye A terminal which has a hole at one end into which a wire is inserted and subsequently soldered.

Solder Eyelet A contact which has a hole into which a wire is inserted and mechanically attached before soldering.

Solder Extraction A technique for unsoldering component leads or wires from holes in a printed wiring board. It consists of a heated tip which melts the solder and a vacuum tube which sucks the molten solder from the hole. Also known as a Solder Sucker.

Solder Fillet A meniscus shaped configuration of solder around a component lead and the land to which it is soldered.

Solder Flux A material which prevents oxidation during heating and also transforms passive surfaces into active, solderable surfaces. See Flux.

Solder Fusion A method which changes electroplated tin-lead on a circuit into a strong bond to the base copper.

Solder Glasses A type of glass that has a low softening point which wets and bonds to metal and ceramic surfaces.

Solder Immersion A test in which the metal leads of an electronic package are immersed in molten solder to determine the resistance to soldering temperatures.

Soldering A process of joining metals by fusion and solidification of an alloy having a melting point of less than 800 degrees Fahrenheit and without melting the base metals.

Solder Land See Pad.

Solder Leaching The dissolving or alloying of a metal to be soldered into the molten solder.

Solderless Connection A connection formed between two metals by pressure only. No heat or solder is used.

Solderless Terminal A connection which is made by securing a wire inside a metal sleeve or similar device with a crimping tool.

Solderless Wrapped Connection The joining of a wire, under tension, around a square or rectangular post or terminal.

Solder Lugs Metal devices which readily accept a wire and are subsequently soldered.

Solder Mask A coating layer applied over selected areas of a circuit board thereby allowing soldering of only the exposed (uncoated) etched circuit areas; that is, the coating acts as a mask to prevent soldering of coated areas.

Solder Oils Special formulated liquids for use in solder pots, wave soldering equipment, etc. which cover the surface of the molten solder to prevent oxidation and the formation of dross.

Solder Paste A homogeneous mixture composed of minute low temperature melting solder particles, solvent, flux, and other agents. It can be applied by dispensing equipment or screened on the substrate. Also known as Solder Cream.

Solder Projection A solder protrusion extending from a solder joint.

Solder Resist A material used to mask off and surface insulate those areas where solder is not wanted, usually around component mounting holes.

Solder Sleeve A heat shrinkable tube containing a predetermined amount of solder and flux used for environmental resistant solder connections and shield terminations.

Solder Spheres Nearly perfect metal spheres from approximately 0.010 inches minimum in diameter (+/- 0.002 inches) to a variety of larger sizes. Spheres are available as pure metals as well as alloy composites. They have a wide range of applications under SMD's and coating thickness are critical in SMT and other metal joining applications.

Solder Splatter Solder fragments, in a variety of configurations.

Solder Webbing A solder bridge between two conductor patterns or circuits which may or may not be bonded to the base laminate and must be removed.

Solder Wicking A flow of solder over the surface of each conductor of stranded wire completely covering each strand and extending under the insulation. This destroys the flexibility of the stranded wire, resulting in brittleness of the wire.

Solid Logic Technology (SLT) The screening and firing of silver palladium conductors on an alumina substrate as was practiced in the early 1960's.

Solid Metal Mask A thin sheet of metal, normally less than 10 mils. with an etched pattern used in contact printing and stenciling.

Solid Phase Bond A bond between two parts in which no liquid phase occurred prior to or during the joining process.

Solid State A technology in which components and circuits are made from semiconductor materials.

Solid Tantalum Chip A chip or leadless capacitor which contains a solid electrolyte (TaO_5) instead of a liquid.

Solidus The highest temperature at which a metal or alloy is completely solid.

Solubility The extent that one material will dissolve in another.

Solvent Any substance, usually a liquid, which dissolves other substances.

Solvent Resistant A material that is not affected by solvents and does not degrade when in contact with solvents.

Source Connector A fiber optics, a connector which interconnects a light source (e.g. LED) to a fiber optic cable.

Source Region Electromagnetic Pulse (SREMP) The EMP from a nuclear blast at or near the surface of the earth.

Space Transformer A package which has been changed from a dense set of chip connections to a less dense set of connections.

Spacing The linear distance between adjacent conductor edges.

Spade Connector A type of terminal which has a slotted tongue and nearly square sides.

Spade Tongue Terminal A terminal which has a slotted area to accommodate a screw or bolt which can be loosened or tightened for the removal or insertion of the terminal without removing the nut.

Spark Gap A device with two electrodes which are separated by a gas or air. After an electrical discharge the insulation (air or gas) is self restoring. It is used as a switching or protective device.

Specification A precise statement of a set of requirements of a material, system, or service and a procedure to determine whether the requirements have been met.

Specific Gravity The weight of a substance divided by the weight of the same volume of water at the same temperature.

Specific Heat The ratio of a material's thermal capacity to that of water at 15 degrees Celsius.

Specimen A sample of material or a device taken from a production lot which is to be used for testing.

Spike A pulse with a very short time and a greater amplitude than the average pulse.

Spinel A single crystal magnesium aluminum oxide substrate

Spiral Flow Test A method for determining the flow properties of a thermosetting resin in which the resin flows along the path of a spiral cavity. The length of the material which flows into the cavity and its weight give a relative indication of the flow properties of the resin.

Split-Tip Electrode Same as Parallel-Gap-Electrode.

Spray-Up Techniques in which a spray gun is used as the processing tool. In reinforced plastics, for example, fibrous glass and resin can be simultaneously deposited in a mold. In essence, roving is fed through a chopper and ejected into a resin stream which is directed at the mold by either of two spray systems.

Spring-Finger Action A design of a contact permitting easy stress-free spring action which provides contact pressure or retention. Used on printed circuit boards and socket contacts.

Sputter Cleaning Prior to evaporating layers of conductive or dielectric layers on a substrate, the surface of the substrate is cleaned by the bombardment with argon or other types of gas ions to remove oxide films. This improves the adhesion of the layers to the surface of the substrate.

Sputtering A thin film process for depositing material on a substrate. The substrate is placed in a vacuum chamber directly opposite a cathode made of the metal or dielectric to be sputtered or evaporated. The cathode is then bombarded with positive ions and small particles of the cathode material deposit uniformly on the substrate.

Squeegee A straight edge rubber blade that pushes the paste/ink across the screen and through the mesh onto the substrate or work piece. Usually made of a rubber having a durometer of 60 - 80.

Stabilizers Chemicals used in plastics formulation to assist in maintaining physical and chemical properties during processing and service life. A specific type of stabilizer, known as an ultraviolet stabilizer, is designed to absorb ultraviolet rays and prevent them from attacking the plastic.

Stainless Steel Screen See Screen.

Stair Step Print
In thick film, a print which resembles the mesh of the screen along the edges of circuitry and resistors. This effect can be attributed to insufficient squeegee pressure or emulsion thickness.

Standard Deviation
A measure of variation of data from the average. It is equal to the root mean square of the individual deviations from the average.

Standard Epoxy Glass
A mixture of approximately 60% epoxy resin and 40% fiber glass.

Standardization
The establishment of common terms, practices, criteria, processes, equipment, parts, assemblies to achieve the maximum uniformity of items, and optimum interchangeability of parts and components.

Standards
References which are used as a basis for comparison or calibration.

Stand-Off
The gap between the substrate/PWB and the bottom mounted component which is mounted on it. The gap is critical in subsequent cleaning processes.

Standoff Insulator
A post made of an insulating material which is used to support a wire above the surface of a structure to which it is mounted.

Static Charge
An electrical charge that clings to the surface of an object.

Static Flex
The installation of the flexible printed wiring assembly to a fixed position in a system.

Steady State
A condition in a circuit in which the voltage, frequency, current, etc. settled down after the initial start up and remain constant thereafter.

Steatite
A ceramic material consisting mainly of silicate of magnesium and used primarily in ceramic substrates because of its insulating properties.

Stencil
A mask which is used to produce an image on the surface of a part.

Step And Repeat
A process in which a circuit pattern is repeated several times to form multiple images. The multiple images are subsequently used to print several substrates or PWB simultaneously.

Step Soldering A technique for making solder connections within an electronic package by using solder alloys with higher melting points initially and subsequently lower melting points solders thereafter.

Stitch Bond (1) A point to point wiring connection system. Insulated wires are welded to hardware which has been inserted into holes in PWB's. (2) A wire bond which is made by laying the wire on a bonding pad and scrubbing with a capillary type bonding tool, by thermocompression or ultrasonic means, to form the joint.

Stop Plate A device attached to a crimping tool which correctly locates a terminal, splice, or contact prior to crimping. Also called a Locator.

Storage Control Element A device which directs the transfer of data and interfaces between the channels, processor storage, and central processor. Also known as the System Control Element.

Storage Hierarchy A combination of the memory elements and their controls. These two form the memory for the processor.

Storage Life See Shelf Life.

Straight Through Lead A component lead which extends through a hole of a printed wiring board and does not require bending.

Strain The deformation resulting from a stress, measured by the ratio of the change to the total value of the dimension in which the change occurred; the unit change, due to force, in the size or shape of a body referred to its original size or shape. Strain is nondimensional but is frequently expressed in inches per inch or centimeters per centimeter, etc.

Stranded Wire A conductor consisting of a group of individual wires which are usually twisted or braided together. This results in a more flexible conductor.

Stratification A layering phenomenon which is caused by the separation of materials in the layers during firing. It is due to the differences in densities of the materials.

Strength, Dry The strength of an adhesive joint determined immediately after drying under specified conditions or after a period of conditioning in the standard laboratory atmosphere.

Strength, Wet The strength of an adhesive joint determined immediately after removal from a liquid in which it has been immersed under specified conditions of time, temperature, and pressure.

Stress The unit force or component of force at a point in a body acting on a plane through the point. Stress is usually expressed in pounds per square inch.

Stress Corrosion A gradual deterioration of the mechanical properties of a material, usually accompanied by crack propagation, caused by the acceleration of applied stress. This normally occurs in a corrosive atmosphere, such as one having high humidity.

Stress Free A material whose molecules are no longer in tension. A material which has been partially annealed or stress relieved.

Stress Relaxation The time-dependent decrease in stress for a specimen constrained in a constant strain condition.

Stress Relieve A process in which the molecules of a material are no longer in stress. This is accomplished by heat cycling the material, such as reheating a film resistor.

Strip To remove the insulation from a wire or cable.

Strip Contacts Contacts which are supplied in a continuous length which are used on automatic equipment.

Stripline (1) A microwave conductor on a substrate. (2) A conductor located between two ground planes such as buried conductors in a multilayer PWB.

Strip Terminal Terminal material supplied in a continuous length which is used on automatic crimping equipment. Also called Tape Terminal.

Stub The wire which connects the inputs of a circuit to the main signal line.

Stud (1) A metal post which is used for connecting wires. (2) A metal post which connects the top side of a printed wiring board to the bottom side or from one level of conductors to another in a multilayer substrate or PWB.

Stylus A needle shaped probe which is used to make electrical contact on a pad of a leadless device or a film circuit.

Subassembly Two or more parts which are combined into a single unit. This is usually done for ease in the assembly and disassembly for servicing.

Subcarrier Substrate A small substrate containing devices and circuits that is mounted on a larger substrate.

Subminiaturization A technique in which miniature components and circuits are used to reduce the volume of electronic packages.

Substrate A base material which provides a supporting surface for deposited or etched wiring patterns.

Subsystem A part of a system which performs part of the system function. It can be removed intact and tested separately.

Subtractive Process A process for obtaining conductive patterns by the selective removal of unwanted portions of a conductive foil.

Superconductor A material which has zero resistance to the flow of electric current.

Supporting Plane A planar structure which is added to a printed wiring board to provide additional mechanical strength, thermal conductivity, dielectric strength, electrical insulation, mechanical stability, etc. It can be added internally or externally.

Surface Creepage Voltage See Creepage.

Surface Diffusion The high temperature injection of atoms into the surface layer of a semiconductor material to form the junctions. Usually a gaseous diffusion process.

Surface Finish The geometric irregularities in the surface of a solid material. Expressed in micro inch per inch.

Surface Leakage The flow of current over the boundary surface of an insulator rather than through its volume.

Surface Mounted Component (Leaded) A discrete component, such as a resistor, capacitor, diode, transistor which is bonded to the surface of conductive patterns on a substrate or printed wiring board without using holes in the substrate or PWB.

Surface Mount Technology (SMT) The technology of mounting components, hybrid circuits, and electronic packages on the surface of printed wiring boards as opposed to inserting leads of components in holes in PWB's.

Surface Preparation A physical and/or chemical preparation of an adherend to render it suitable for adhesive joining.

Surface Resistivity The resistance of a material between two opposite edges of a unit square of its surface.

Surface Tension The tendency of the surface of a liquid to contract which is caused by the intermolecular attraction of the molecules below the surface of the liquid.

Surface Texture The smoothness or roughness of the surface of a material. Usually a surface finish requirement is specified in the drawing.

Surfactant A material which improves the wetting of the surface of a material.

Surge A sudden and/or abrupt change in voltage and current in a circuit.

Surge Protective Device A device that reduces or diverts surges of currents in a circuit.

Surge Withstand Capability The ability of a circuit or an electrical system to avoid damage to its components because of excessive voltages.

Susceptibility A condition of a system whereby it is easily affected by interferences, transients, and signals other than those to which the equipment was designed.

Swaged Leads Component leads which have been deformed and secured on the noncomponent side of a printed circuit board for subsequent soldering to the board. Same as Swaging or Swedging.

Swim The floating of components in the X-Y directions during reflow soldering.

Swimming In thick films, the lateral moving of a thick film conductor pattern on molten glass crossover patterns

Switching Noise A noise which is created by an induced voltage at the circuit terminals. This induced voltage is created by the rapidly changing current caused by the switching of many devices.

Syntactic Foams Light-weight systems obtained by the incorporation of prefoamed or low-density fillers in the systems.

System A group of interconnected parts and devices designed for a specific objective by performing certain functions.

Tab See Printed Contact.

Tack The property of an adhesive that enables it to form a bond of measurable strength immediately after adhesive and adherend are brought into contact under low pressure.

Tail of the Bond In wire bonding, the end of the wire which extends beyond the wire bond from the heel. The tails are removed later after the wire bonding is completed.

Tail Pull In wire bonding, the removing of the tails from the wire bonds.

Tantalum Capacitor An electrolytic capacitor which uses tantalum foil or a sintered slug of tantalum as the anode. The tantalum oxidizes and forms the dielectric barrier.

Tape, Alumina Substrates are made by casting alumina slurries to a predetermined thickness and diced. This green tape or soft substrate material is then cut to size, holes punched, and subsequently fired. Also known as Green Tape.

Tape Automated Bonding (TAB) A process in which precisely etched leads, which are supported on a flexible tape or plastic carrier, are automatically positioned over the bonding pads on a chip and a heated pressure head is lowered over the assembly and simultaneously thermocompression bonding the leads to all the pads on the chip.

Tape Terminal See Strip Terminal.

T Dimension See "G" dimension.

Tear Strength Measurement of the amount of force needed to tear a solid material that has been nicked on one edge and then subjected to a pulling stress. Measured in lb/in.

Temperature Aging The stressing of a material, component, or system at an elevated temperature for an extended period of time.

Temperature Coefficient of Capacitance The amount of change in capacitance of any electronic device with respect to temperature. Measured in parts per million per degree Celsius over a specific temperature range.

Temperature Coefficient of Linear Expansion The amount of change in any linear dimension of a solid from a change in temperature. Measured in microinches per inch per degree Celsius. Also called CTE, Coefficient of Thermal Expansion.

Temperature Coefficient of Resistance (TCR) The maximum change in resistance per unit change in temperature, usually expressed in parts per million (ppm) per degree centigrade, and specified over a temperature range. The temperature is that of the resistor not the ambient temperature.

Temperature Cycling An environmental test in which a material, component, or system is subjected to temperature changes. Usually from a low temperature (-55 degree Celsius) to a high temperature (+125 degree Celsius) over a period of time and for a specific number of cycles (10,25,100 cycles).

Temperature Excursion The temperature differences (from the lowest temperature to the highest temperature) which a component, material, or system experiences.

Tensile Strength (1) The maximum tensile stress a material is capable of sustaining. Tensile strength is calculated from the maximum load during a tension test carried to rupture and the original cross-sectional area of the specimen. (2) The maximum amount of axial load required to pull a wire from a crimped barrel of a terminal.

Terminal A metallic pin or device to which an electrical connection can be made. Some of the most common types of terminals are: solder contact, clip, package lead, solderless wrap, spade, ring, flag, blade, flanged and offset.

Terminal Area Same an Terminal Pad or Land.

Terminal Block A block of insulating material on which many terminal connectors have been mounted for making electrical connections.

Terminal Board A board made of an insulating material which has a single or double row of termination pads for making electrical connections to a mating connector.

Terminal Pad Same as Terminal Area or Land.

Termination See End Termination.

Testability The degree or ease to which an electronic circuit or package can be electrically tested.

Test Coupon A sample or test pattern usually made as an integral part of the printed board, on which electrical, environmental and microsectioning tests may be made to evaluate board design or process control without destroying the basic board.

Thermal A generic term relating to heat and all forms of heat type devices such as thermometers, thermocouples, thermopiles, etc.

Thermal Conductivity The ability of a material to conduct heat; the physical constant for the quantity of heat that passes through a unit cube of a material in a unit of time when the differences in temperature of the two faces is one degree Celsius. The rate at which a material conducts heat is expressed in calories per square centimeter per centimeter per cross section per degree Celsius. Rates of thermal conductivity of some materials are:

CRS(C1010)	.12
Alloy 52	.04
Kovar	.0395
Corning 7052 Glass	.0028
Alumina (96%)	.49 at 20 degrees Celsius
Beryllia (98%)	.37 at 100 degrees Celsius
Epoxy	.0003-.0006
Alumina filled epoxy	.0009-.0011
Beryllia filled epoxy	.021-.073
Teflon	.0006
Aluminum	.48
Copper	.918

Thermal Conduction Module A module containing 100 chips or more which is cooled by thermal conduction methods in contact with the chips.

Thermal Cycling A method used to induce a stress on electrical components by heating and cooling in an air circulating oven to accelerate reliability testing.

Thermal Design A study of the flow of heat and their paths from each heat dissipating component in a circuit to an external heat sink.

Thermal Drift The change from the nominal values of components and circuit elements due to changes in temperature.

Thermal Drop The difference in temperature across a boundary or across a material.

Thermal Expansion (CTE) Coefficient of thermal expansion. Measurement of the rate at which a given material expands as heat increases, expressed as a factor of 10^{-7} cmper degree Celsius.

C-1010	140
Kovar	55
52 Alloy	90
Glass	55
Alumina	50
Beryllia	70
Copper	160
Stainless Steel	160

Thermal Gradient The variation in temperature across the surface of or through a material being heated.

Thermal Mismatch The differences of the CTE of two materials which are joined together.

Thermal Network A model of a total system which is broken down into subsystems, each displaying its thermal property and connected to the other members of the subsystem so as not to distort the total thermal property of the system.

Thermal Resistance The resistance offered by a material or medium to the flow of thermal energy through the medium.

Thermal Shock An environmental test in which devices or equipment are subjected alternately to high and low temperatures rapidly in attempting to cause a failure.

Thermal Stability The resistance of a material to changes in physical, electrical, and chemical properties caused by heat.

Thermal Stress Cracking Crazing and cracking of some thermoplastic resins which result from overexposure to elevated temperatures.

Thermocompression Bonding The joining of two materials by interdiffusion across the boundary by the application of heat and pressure.

Thermoforming A process in which thermoplastic resins in sheet form are heated to their softening points, drawn or forced into an open mold and subsequently cooled.

Thermogravimetric Analysis (TGA)
A method which measures the change in weight of a material as a function of increasing the temperature of the material.

Thermomechanical Analysis (TMA)
A method which measures the linear expansion or contraction of a material as the temperature of the material is increased or decreased.

Thermoplastic
A plastic which is set into its final shape by forcing the melted base polymer into a cooled mold or through a die after which it is cooled. The hardened plastic can be remelted and reprocessed several times. Thermoplastics are also known as linear or branched polymers.

Thermosetting
A plastic which is cured, set or hardened, usually by heating, into a permanent shape. The polymerization reaction is an irreversible reaction known as crosslinking. Once set, a thermosetting plastic cannot be remelted, although most will soften with the application of heat.

Thermosonic Bonding (T/S)
A bonding process which combines thermocompression and ultrasonic bonding. A wire is bonded by applying ultrasonic power to the capillary of a thermocompression bonder.

Thevenin Equivalent
A means to better understand the voltage-current behavior in an electrical system between two of its nodes. It can be a constant voltage source with a series impulses or a constant current source shunted by an impedance.

Thick Film
A layer of conductive, resistive, or dielectric ink or paste silk-screened onto a substrate which is heated to form conductors, resistors, and capacitors. Heating is known as _firing_ for cermet thick films or _curing_ for polymer thick films. Layers are typically about 0.001 inches thick.

Thick-Film Circuit
A microcircuit, including passive devices such as resistors and capacitors, which are screen printed on a ceramic substrate and fired.

Thick Film Dielectric
A screen printable paste composed of finely ground insulating materials such as glass or ceramic powders.

Thick-Film Hybrid Circuit
A thick film circuit on a substrate to which chip devices have been added.

Thick-Film Network
A network of screen printed resistors and capacitors that are interconnected by screen printed conductors on a ceramic substrate and subsequently fired.

Thick Film Resistor, Conductor, And Dielectric Compositions
Screen printable pastes composed of metals, oxides and glass powders.

Thick Film Technology A technology in which electronic circuits and networks are formed by screen printing conductive, resistive, and dielectric layers on a ceramic substrate and firing. Conductive layers are approximately 0.0003 - 0.001 inches thick; dielectric 0.001 - 0.002 inches and resistors 0.0004 -0.0008 inches.

Thin Film A layer of conductive, resistive, and dielectric material sputtered or evaporated onto a substrate in a vacuum to form conductors, resistors, and capacitors. Layers are typically less than 1,000 angstroms thick. (1 angstrom is approximately 4×10^{-9} in. or 10^{-8} cm.).

Thin Film Capacitor A capacitor which is made by the evaporation or sputtering of conductive and dielectric materials in the form of layers.

Thin Film Circuit A circuit in which active or passive devices and conductors are produced as films on a substrate by the evaporation of dielectric, resistive, and conductive materials.

Thin Film Hybrid Circuit A thin film circuit on a substrate to which chip devices have been added.

Thin Film Integrated Circuit A microcircuit, including passive devices, that are produced as films by evaporation or sputtering techniques on a substrate.

Thin Film Network A network of resistors and or capacitors and interconnected by conductors on a substrate which are vacuum deposited by evaporation or sputtering techniques.

Thin Film Packaging Electronic packages whose substrates were fabricated using thin film conductors and insulators.

Thin Film Technology A technology in which electronic circuits and networks are formed by vacuum evaporation or sputtering techniques on a substrate. The films range in thickness from 0.3 - 1.0 micro meters.

Thinner A volatile liquid added to an adhesive or other resin forms to modify the viscosity of the material or compound.

Thixotropic Describing materials that are gel-like at rest but fluid when agitated.

Three Layer Tape In tape automated bonding (TAB) a tape which contains two metallized layers with a dielectric layer in between and the metal layers interconnected.

Through Connection An electrical connection between conductors on opposite sides of a substrate or base material.

Through Hole (Plated) A hole which is drilled or punched in a printed circuit board and subsequently metallized thereby electrically connecting the two sides of the board.

Through-Hole Component A leaded component designed for mounting on PWB's by inserting the leads through the holes and subsequently soldering them to circuitry on the board.

Time, Assembly The time interval between the spreading of the adhesive on the adherend and the application of pressure or heat, or both, to the assembly.

Time, Curing The period of time during which an assembly is subjected to heat or pressure, or both, to cure the adhesive.

Time Domain Reflection A rise and fall in current and voltage at the point of a break in a transmission line creates a disturbance or reflection which travels in the opposite direction which causes line noises and signal distortions.

Tin (Sn) A widely used metal for plating brass, copper and steel terminals. Excellent electrical and thermal conductivity and used on component leads for solderability and to reduce galvanic corrosion when in contact with aluminum.

Tin-Lead (Sn-Pb) An alloy, usually 62% tin and 38% lead used in most soldering applications because of its low melting point.

Tinning To coat metallic surfaces with a thin layer of tin or solder.

Tip In wire bonding, that part of the bonding tool which applies pressure to the wire to form the bond.

Toe See Tail of the Bond.

TO Can Abbreviation for Transistor-Outline Metal Can Package.

T O Package Abbreviation for Transistor Outline Package. Established by JEDEC as an industry standard. A cylindrical shaped metal package containing and IC chip. The chip is eutectically mounted on a base and interconnected to terminals with feed through glass to metal seals. A metal cover, in the form of a top hat, is placed over the IC and hermetically sealed to the metal base.

Tombstone The raising-up of small chips from the substrate during solder reflow which is caused by surface tension and unbalanced forces of solder wetting. This could result in an open wire bond or solder termination. The preferred term for Drawbridging, Manhattan Effect.

Top Hat Resistor A thick film resistor which has a projection extending from one side which can be trimmed, in addition to the main body of the resistor, thereby increasing the resistivity.

Topography In thick film technology, it is the condition of a fired surface with respect to valleys and bumps.

Topology A surface layout of the elements of an integrated circuit.

Top Side Metallurgy The metallized pads on the surface of a substrate to which chips are attached or electrically connected.

Toroids A ring or doughnut shaped coil winding which is used for inductors and transformers because of its volumetric efficiency.

Track (1) A path of deteriorated material on the surface of a dielectric. (2) A synonym for a conductor on a substrate.

Tracking A path of conductive or nonconductive contaminate left on the surface of an insulating material after an arcing condition occurs.

Track Resistance The resistance of organic materials, such as plastics, to the formation of a carbon track on the surface of the material by high voltage arcing. Sometimes also called arc resistance. Correctly defined, arcing is the electrical condition and tracking is the carbon path resulting from deterioration of the material by the arcing condition. (See Arc Resistance.)

Transfer Molding A method of molding thermosetting materials, in which the plastic is first softened by heat and pressure in a transfer chamber, then forced by relatively low pressure through suitable sprues, runners, and gates into a closed mold for final curing. Electronic assemblies can be safely molded by this process due to the relatively low molding pressures.

Transient A sudden change in conditions in a system which last for a short period of time.

Transient Control Level The highest peak voltage which a system can take without causing damage to the circuit or system.

Transient Mismatch A difference in thermal conductivity of two elements in a system and because of this difference a lag occurs until equilibrium is reached.

Transient Radiation Effects On Electronics (TREE) The damage incurred to the electronics of a system caused by photons or subatomic particles.

Transistor An active semiconductor device having three or more electrodes. It is usually made of germanium or silicon and used as an amplifier, detector or switch.

Transistor Outline Package See TO Package.

Transistor-Transistor Logic (TTL) A widely used, relatively high speed, medium power dissipation, inexpensive form of semiconductor logic. Its basic logic element is a multiple emitter transistor.

Translam-Transfer Lamination An additive plating and laminating process for producing fineline printed circuit boards. Rapid Impingement Speed Placing (RISP) is used as the plating process. Each layer is plated on a stainless steel plate, covered with B-stage dielectric material and transferred to a board stack-up for final lamination.

Transmission Line (1) A conductor or a series of conductors used for carrying electrical energy from a source to a load. (2) A circuit used for transmitting pulse signals with controlled electrical characteristics.

Transmission Loss The amount of power lost by a signal as it is transmitted from one point to another.

Trimming The process of using abrasive or laser techniques to accurately remove thick-film material from a substrate to change (increase) the resistance value.

Trimming, Functional The process of adjusting (usually increasing the values of a resistor) on an operating circuit to a specific voltage or current at the output.

Trim Notch See Kerf.

Twill Weave A basic weave characterized by a diagonal rib, or twill line. Each end floats over at least two consecutive picks enabling a greater number of yarns per unit area than a plain weave, while not losing a great deal of fabric stability.

Twisted Pair Two insulated wires which are twisted around each other.

Two Layer Tape In TAB, a dielectric tape, Kapton, containing one metallized layer on one side of the tape and etched to form the TAB leadframe.

Type I SMT A SMA which has 100% surface mounted components and all the connections are made by solder reflow processes. Type I includes both single and double sided boards.

Type II SMT A SMA which has a mixture of SMC's and through hole components on the same side of the PWB and the connections are made by solder reflow and flow soldering processes. Type II can be a double sided PWB but one side must have a mixture of components on one side.

Type III SMT A SMA which has IMC's on the component side of the PWB and SMC's adhesively bonded to the bottom side of the board. All the solder connections are made simultaneously by solder reflow processes with the bonded SMC's on the underside of the board.

Ultra High Frequency (UHF) In the radio frequency spectrum, the band extending from 300 to 3,000 MHz.

Ultrasonic Bonding A metal joining process in which two metals are bonded together by applying pressure plus an ultrasonically induced scrubbing action to form a molecular bond.

Ultrasonic Cleaning A method which utilizes the cavitation of a chemical solvent to aid in the cleaning. The cavitation is applied by ultrasonic energy from a high frequency generator and a transducer.

Ultra Large Scale Integration (ULSI) A chip which contains a minimum of 100,000,000 transistors.

Ultraviolet Cure A process in which crosslinking of certain polymer materials, such as epoxy adhesives, Film emulsions, etc., is activated by the exposure to UV radiation.

Ultraviolet Rays (UV) Light waves having wave lengths from 200 - 4,000 angstroms.

Underbond Insufficient deformation of the wire by the bonding tool in the wire bonding operation.

Undercut In the chemical etching of conductors, it is the removal of the metal conductor from under the edge of the resist. Also called Overetching.

Underglaze In thick film technology, it is the application of a glaze material to a substrate, where the resistors are to be placed, prior to the screening of resistors.

Universal Pattern A circuit board pattern or patterns which will accept standard size packages and configurations such as dual-in-line packages.

Ureas A thermosetting resin characterized by good chemical and electrical properties. Also known for their hard scratch resistant finish and high arc resistance as well as resistance to heat. (Up to 170 degrees Fahrenheit.

UV Stabilizer (Ultraviolet) Any chemical compound which, when admixed with a thermoplastic resin, selectively absorbs ultraviolet rays.

Vacuum-Bag Molding A process for molding reinforced plastics in which a sheet of flexible transparent material is placed over the lay-up on the mold and sealed. A vacuum is applied between the sheet and the lay-up. The entrapped air is mechanically worked out of the lay-up and removed by the vacuum, and the part is cured.

Vacuum Bake A process in which unsealed packages, printed wiring boards, etc. are placed in a vacuum chamber and heated to temperatures up to 150 degrees Celsius for periods of time, from several hours to several days, while constantly being pumped down to vacuum pressures of 2 - 5 m.m. to remove moisture and other contaminants from the packages and boards.

Vacuum Deposition The deposition of thin metal/dielectric films on a substrate by the evaporation of the materials in a vacuum chamber.

Vacuum Evaporation The vaporizing by heating of metal and dielectric materials at reduced pressures. These materials are deposited as thin films on a substrate.

Vacuum Injection Molding Molding process using a male and female mold wherein reinforcements are placed in the mold, a vacuum is applied, and a room-temperature-curing liquid resin is introduced which saturates the reinforcement.

Vacuum Pickup A tool used for picking up chip devices. It consists of a pencil shaped tube, with a non scratching tip, at the pick up end, while the other end is connected to a low vacuum source.

Vapor Deposition See Vacuum Deposition.

Vapor Phase Reflow Same as Vapor Phase Soldering.

Vapor Phase Soldering A method for soldering many solder joints simultaneously. The process is performed in a heated chamber containing boiling fluorinated hydrocarbons with two separate vapor phase zones. For example, Freon TF and FC-5312 which have vapor phase zones of 185 - 215 degrees Fahrenheit and 419 degrees Fahrenheit respectively. Parts and assemblies are heated by the heat of condensation as the cold parts are introduced into the hot vapor.

Varistor A semiconductor, nonlinear device whose resistance drops as the applied voltage is increased.

Varnish A liquid resin which is used to coat electrical components and impregnate electrical coils to provide electrical, mechanical and environmental protection.

Vehicle A liquid which is added to thick film paste or other viscous or paste materials to adjust the viscosity.

Veil Coat A resin enriched surface next to the mold which provides a smooth surface over the coarse fibers of the reinforcement.

Very High Frequency (VHF) In the radio frequency spectrum, the band extending from 30 to 300 MHz.

Very High Speed Integrated Circuit (VHSIC) A Department of Defense directed program to develop very high speed IC's with device rise times ranging from approximately one nanosecond to subnanoseconds.

Very Large Scale Integration (VLSI) An IC chip which contains a minimum of 10,000 transistors.

Very Low Frequency (VLF) In the radio frequency spectrum, the band extending from 10 to 30 KHz.

Vessication See Mealing.

Via An opening in a dielectric layer which is filled with a conductive paste to form an interconnection between multilayers of a thick film hybrid circuit.

Via, Blind A via which extends from the surface of a substrate to one of the inner layers.

Via Buried A via which connects inner layer circuits but does not extend to the surface of a thick film substrate or multilayer printed circuit board.

Via, Fixed A via which exists on a prescribed grid pattern on the various layers of a thick film multilayer substrate and can connect adjacent layers and nonadjacent layers.

Via Hole A plated-through-hole but not used for the insertion of a component lead. Its sole purpose is for interconnecting two or more internal layers.

Via, Programmable A via which is not on a prescribed grid pattern as the fixed via.

Via, Thermal See Through Via.

Via, Through A fixed via which extends through all of the metallized layers of a thick film multilayer substrate. It can also be used for electrical grounding or thermal dissipation. Also know as a stacked or thermal via.

Via, Stacked See Through Via.

Vicat Softening Temperature A temperature at which a specified needle point will penetrate a material under specified test conditions.

Viscoelastic A characteristic mechanical behavior of some materials which is a combination of viscous and elastic behaviors.

Viscometer An instrument capable of measuring the viscosity of fluids and pastes.

Viscosity A measure of the resistance of a fluid to flow in a simple liquid or Newtonian fluid. The unit of measurement is the pascal second (Pa.s). Frequently the centipoise (CP), which is one millipascal (m Pa.S), is used as the viscosity unit. For simple liquids, the viscosity is constant at all shear rates. In thick film pastes, a non Newtonian liquid, the viscosity will vary depending on the shear rate.

Vitreous A term meaning glassy or glass like.

Vitreous Binder A glassy material which is added to thick film paste to bind all the particles together. Dielectric paste are nearly 100% glass materials. A conductor paste used for inner layers contain 5 - 10% glass while the outer layers are fritted and contain less than 1%.

Vitrification The conversion into a glass like material. This reduces the porosity of a material and results in lower moisture absorption.

Void A small hole or space in a localized area of a solid material.

Voltage Breakdown The voltage required to cause an insulation failure.

Voltage Endurance The extended period of time in which it takes an insulating material to fail a prolonged voltage stress.

Voltage Regulation A measure of the ability of a voltage regulation device to maintain its output voltage under varying load conditions.

Volume Resistivity (Specific Insulation Resistance) A measure of the The electrical resistance between opposite faces of a 1 cm cube of insulating material, commonly expressed in ohm-centimeters. The recommended test is ASTM D 257-54T.

Vulcanization A chemical reaction in which the physical properties of an elastomer are changed by causing it to react with sulfur or other cross-linking agents.

Vulcanized Fiber A cellulosic material which has been partially gelatinized by action of a chemical (usually zinc chloride), then heavily compressed or rolled to a required thickness, leached free from the gelatinizing agent, and dried.

Vulnerability An undesirable condition in which disturbances can damage a circuit or system.

Wafer A thin disc of semiconducting material on which many chips are fabricated at one time. The chips are subsequently scribed, separated and packaged individually.

Waffle Pack A flat package containing rectangular shaped cavities for the storage and protection of bare IC devices. It resembles an egg crate configuration and has a cover with locking devices an the sides. Also called Matrix Tray.

Warpage Dimensional distortion in a plastic object after molding.

Warp And Woof The threads or wires in a woven screen which cross over and under each other at right angles.

Water Absorption The ratio of the weight of water absorbed by a material to the weight of the dry material.

Water-Extended Polyester A casting formulation in which water is suspended in the polyester resin.

Wave Soldering A process in which many potential solder joints are brought in contact with a wave of flowing solder for a short period of time and the joints soldered simultaneously.

Wear Out That span of time beyond the failure rate period in which the failure rate of the component exceeds specific predicted values.

Weave Exposure A condition in which the unbroken woven glass cloth is not uniformly covered by resin.

Weave Texture A surface condition in which the unbroken fibers are completely covered with resin but exhibit the definite weave pattern of the glass cloth.

Webbing, Solder Solder residue which adheres to the laminate material (e.g., epoxy glass) in threadlike form.

Wedge Bond A wire bond which is made with a wedge shaped tool. Most ultrasonic bonds are wedge type bonds.

Welding A process in which two metals are joined together by the application of heat, causing the metals to melt and fuse together. Welding can be accomplished with or without a filler material.

Wettability, Soldering The degree to which a metal surface to be soldered is clean (free of oxides) and metal contact can be made between the solder and the metal to be soldered.

Wetting (1) Ability to adhere to a surface immediately upon contact. (2) In soldering, the ability of molten solder to spread over a metal surface after the application of a flux and the proper amount of heat.

Whisker A metallic growth needle-like in size and shape, which appears on the surface of a printed wiring board and is caused by components whose leads have been electroplated with tin, the presence of moisture, and some voltage potential.

Wicking Another method of desoldering component leads or wires from a metallized hole. A prefluxed braid of stranded wire is placed on the solder joint and heated with the tip of a soldering iron. The solder is removed from the hole by capillary action wicking into the braided wire.

Wire A solid or a stranded group of metal conductors. It can be round, square, or rectangular in shape, either bare or insulated and have a low resistance to the flow of current.

Wireability The degree to which a package (e.g., a printed wiring assembly) allows the interconnection of components and hybrid packages and terminals mounted on it as to the probability of successful interconnection. The wireability of a package is acceptable when there exits additional wiring capacity, via availability, and access to terminals.

Wirebond A wire connection between a pad on semiconductor chip and a pad on a substrate. It consists of pads on both the die and the substrate, the fine wire, and the interfaces between the wire and metal surfaces on the die and substrate.

Wire Bonding A method used to attach a fine wire, usually 1 mil. in diameter, to pads on substrates, and other bonding surfaces. Several methods such as thermocompression, ultrasonic, and thermosonic are presently used in wirebonding.

Wire Length,Average The average length of wire of all the connections in a specific package level.

Wire Sag The failure of bonding wire to form the loop defined by the path of the bonding tool between bonds.

Wire Wrapped Connection Same as Solderless Wrapped Connection.

Wiring A manual or automatic process of interconnecting components or chips with wires.

Wiring Overflow See Overflow Wiring.

Working Life The period of time during which a liquid resin or adhesive, after mixing with catalyst, solvent, or other compounding ingredients, remains usable. Similar to Pot-Life.

Woven Screen A screen having a specific mesh size which is used for screen printing. Screens can be of various mesh sizes and made from stainless steel, silk, or nylon.

Woven Roving A heavy glass-fiber fabric made by the weaving of roving.

Wrapped Connection Same as Solderless Wrapped Connection.

Yield Value The lowest stress at which a material undergoes plastic deformation. Below this stress, the material is elastic, above it, viscous. Also, the stress at which a material exhibits a specified limiting deviation from the proportionality of stress to strain. Same as Yield Strength.

Zener Diode A two layer semiconductor diode designed to be operated in the reversed-biased breakdown condition.

Zero Insertion Of a connector. A type of connector in which all the electrical contacts touch simultaneously, with no insertion force required, but only after the connector halves are engaged and properly aligned.

3-D Circuit Same as Molded Circuit.

ABBREVIATIONS, ACRONYMS, AND SYMBOLS

A

ABS	Acrylonitrile-Butadiene-Styrene
ACPI	Automated Component Placement And Insertion
AgPd	Silver Palladium
AI	Artificial Intelligence
AIA	Aircraft Industries Association
AID	Automatic Insertion Dip
ALN	Aluminum Nitride
ALU	Arithmetic Logic Unit
Al_2O_3	Alumina
ANSI	American National Standards Institute
AP	Arithmetic Processor
APC	Array Processor Controller
APE	Asynchronous Processing Element
APIO	Array Processor Input/Output
AQL	Acceptable Quality Level
As	Arsenic
ASIC	Application Specific Integrated Circuits
ASP	Advanced Signed Processor
ASTM	American Society For Testing And Materials
ASW	Anti Submarine Warfare
ATAB	Area-Array Tape Automatic Binding
ATE	Automated Test Equipment
AuGe	Gold-Germanium
AuPt	Gold-Platinum
AuSn	Gold-Tin
AuSi	Gold-Silicon
AWG	American Wire Gauge

$BaTiO_3$ Barium-Titanate

BeO   Beryllia. Beryllium Oxide (Bromellite)

BIT   Built In Test

BMC   Bulk Molding Compound

BN   Boron Nitride

BOPS   Billion Operations Per Second

Brassboard   Field Demonstratable Electronic Module

Breadboard   Laboratory Demonstratable Electronic Module

BTAB   Bumped Tape Automatic Bonding

C   Capacitance; Centigrade

CAD   Computer Aided Design

CAE   Computer Aided Engineering

CAM   Computer-Assisted Manufacturing

CAT   Computer Aided Testing

CAVP   Complex Arithmetic Vector Processor

CC   Chip Carrier

CCC   Ceramic Chip Carrier

CDA   Clean Dry Air

CDR   Critical Design Review

CERDIP   Ceramic Dual In-Line Package

CFC   Chlorinated Fluorocarbon

CGA   Configerable Gate Array

CIM   Computer Integrated Manufacturing

CLA   Centerline Average (A Measure Of Substrate Surface Roughness)

CLCC   Ceramic Leaded Chip Carrier

CLDCC   Ceramic Leaded Chip Carrier (Preferred)

CLLCC   Ceramic Leadless Chip Carrier (Preferred)

	C

CML — Current Mode Logic

CMOS — Complementary Metal-Oxide Semiconductor

COB — Chip On Board

CPU — Central Processing Unit

CQFP — Ceramic Quad Flat Pack

CSA — Canadian Standards Associations

CTE — Coefficient Of Thermal Expansion

CVD — Chemical Vapor Deposition

C and W — Chip and Wire

	D

DAP — Diallyl Phthalate

Db — Decibel

DI — Deionized Water

DCAS — Defense Contracts Administration Services

DIP — Dual In-Line Package

DMA — Dynamic Mechanical Analysis

DODISS — Department Of Defense Index Of Specifications and Standards

DPA — Destructive Physical Analysis

DRAM — Dynamic RAM

DSC — Differential Scanning Calorimetry

DTA — Differential Thermal Analysis

DUT — Device Under Test

	E

E — Voltage or Potential

ECL — Emitter Coupled Logic

ECM — Electronic Countermeasures

EDS — Energy Dispersive Spectroscopy

EEPROM	Electrically Erasable Programmable Read-Only Memory
EIA	Electronics Industries Association
EIAJ	Electronic Industries Association Of Japan
EMC	Electromagnetic Compatibility
EMI	Electromagnetic Interference
EMP	Electromagnetic Pulse; Electromagnetic Potential
EMPF	Electronic Manufacturing Productivity Facility
EO	Electro-Optic
EOS	Electrical Overstress
EPROM	Electrically Programmable Read Only Memory
ESCA	Electron Spectroscopy For Chemical Analysis
ESD	Electrostatic Discharge
ETPC	Electrolytic Tough Pitch Copper.
EUT	Equipment Under Test
EW	Electronic Warfare

f	Frequency
FA	Failure Analysis
FACI	First Article Configuration Inspection
FCFC	Flat Conductor Flat Cable
FEP	Fluorinated Ethylene Propylene (Teflon)
FET	Field Effect Transistor
FLIR	Forward Looking Infrared
FPAP	Floating Point Arithmetic Processor
FR	Failure Rate
FR-1	Flammability Rating
FRP	Fiber Glass Reinforced Plastics

GaAs	Gallium Arsenide
Ge	Germanium
GFE	Government Furnished Equipment
GHz	Giga Hertz
gnd	Ground

HAZMAT	Hazardous Material
HCC	Hermetic Chip Carrier
HCMOS	High-Density CMOS
HDCM	High-Density Ceramic Module
HDPE	High-Density Polyethylene
HEMP	High Attitude Electromagnetic Pulse
HF	High Frequency
HIC	Hybrid Integrated Circuit
HMOS	High Performance MOS
HOL	Higher Order Language
HTRB	High Temperature Reverse Bias
Hz	Hertz or Cycles Per Second (One Hertz is one cycle per second)

I	Current
IAPU	Image Array Processing Unit
IC	Integrated Circuit; Internal Connection
IDC	Insulation Displacement Connector; Insulation Displacement Contact
IEC	International Electrotechnical Commission
IEEE	Institute Of Electrical And Electronics Engineers
ILB	Inner Layer Bond

IMC	Insertion Mounted Component
I/O	Input/Output
IPC	Institute For Interconnecting And Packaging Electronic Circuits
IPS	Instructions Per Second
IR	Infrared
IRS	Infrared Scan
ISA	Instruction Set Architecture; Imaging Sensor Autoprocessor
ISHM	International Society of Hybrid Microelectronics
I/SMT	Interconnect/Surface Mount Technology
ISO	International Standards Organization

JC	JEDEC Committee
JEDEC	Joint Electronic Devices Engineering Council

K	Symbol For Dielectric Constant; also used for abbreviation of Thousands, 1×10^3.

L	Inductance
LCC	Leadless Chip Carrier
LCCC	Leadless Ceramic Chip Carrier
LCD	Liquid Crystal Display
LCP	Liquid Crystal Polymer
LED	Light Emitting Diode
LEMP	Lightning Electromagnetic Pulse
LDPE	Low Density Polyethylene
LIC	Linear IC

LID	Leadless Inverted Device
LLCC	Leadless Chip Carrier (Preferred)
LMCH	Leadless Multiple-Chip Hybrid
LPGA	Leadless Pad Grid Array
LSI	Larger Scale Integration
LTPD	Lot Tolerance Percent Defective

M	Abbreviation for mega or one million
m	Abbreviation for milli or one thousandth
MCC	Multiple Chip Carrier; Miniature Chip Carrier
MCP	Multichip Package
MELF	Metal Electrode Face (Bonded)
MESFET	Metal-Semiconductor Field Effect Transistor
MFD	Microelectronic Functional Device
MHz	Megahertz (one million cycles per second; i.e., 10 MHz =10,000,000 Hz)
MIC	Monolithic IC; Microwave IC
MICROMETER	Micron = 10^{-6} meters
MICRON	Micrometer = 10^{-6} meters
MIL-STD	Military Standard
MIPS	Million Instructions Per Second
MIS	Metal Insulator Semiconductor
MLB	Multilayer Board
MLC	Multilayer Ceramic; Monolithic Ceramic Capacitor
MMIC	Monolithic Microwave IC
MNS	Metal Nitride Semiconductor
MNOS	Metal Nitride-Oxide Semiconductor
MOPS	Million Operations Per Second

MOS	Metal-Oxide Semiconductor
MOS FET	Metal-Oxide Semiconductor Field Effect Transistor
MRB	Material Review Board
MSI	Medium Scale Integration
MTBF	Mean-Time-Between-Failures; Mean-Time-Between-Fault
MTM	Multiple Termination Module
MTNS	Metal Thick Nitride Semiconductor
MTOS	Metal Thick Oxide Semiconductor
MTTF	Mean Time To Failure
MTTR	Mean Time To Repair

n	nano [1 nano second = (.0000000001 seconds); (10^{-9})]
NDRO	Non-Destructive Read Out
NDT	Non-Destructive Test
NEMA	National Electrical Manufacturers Association
NEMP	Nuclear Electromagnetic Pulse
NMOS	N-Channel Metal-Oxide Silicon

OLB	Outer Lead Bond(er)
OPS	Operations Per Second

p	Pico [1 pico second = 1 millionth of one second) (10^{-12}]
PCB	Printed Circuit Board
PCTFE	Polychlorotrifluoroethylene
PDA	Percent Defective Allowable
PEL	Picture Element In Display

PGA	Pin Grid Array
PIN	P-N Junction With Isolation Region (Diode)
PLCC	Plastic Leaded Chip Carrier
PLDCC	Plastic Leaded Chip Carrier (Preferred)
PPM	Parts Per Million
POS	Porcelain-On-Steel (Substrate)
PQFP	Plastic Quad Flat Pack
PROM	Programmable Read-Only Memory
PSG	Phospho-Silicate Glass
PSP	Programmable Signal Processor
PTF	Polymer Thick Film
PTFE	Polytetrafluoroethylene
PTH	Plated Through Hole
PVC	Polyvinyl Chloride
PVF_2	Polyvinylidene Fluoride
PWB	Printed Wiring Board

QPL	Qualified Parts List

R	Resistance
RAM	Random Access Memory
RF	Radio Frequency
RFI	Radio Frequency Interference
RGA	Residual Gas Analysis
RISC	Reduced Instructions Set Computering
RISP	Rapid Impingement Speed Plating
RMS	Root Mean Square

ROM	Read Only Memory
RTI	Radiation Transfer Index
RTL	Resistor Transistor Logic
RTV	Room Temperature Vulcanizing

SAM	Scanning Acoustic Microscope
SAW	Surface Acoustic Wave
SCC	Stress-Corrosion Cracking
SDI	Strategic Defense Initiative
SEM	Standard Electronic Modules; Scanning Electron Microscope
Si	Silicon
SIMS	Secondary Ion Mass Spectrometry
SIP	Single In Line Package
SLAM	Single Layer Alumina Metallized; Scanning Laser Acoustic Microscope
SLC	Single-Layer Ceramic
SMA	Surface Mounted Assembly
SMC	Surface Mounted Component
SMD	Surface Mounted Device
SMT	Surface Mount Technology
SO	Small Outline
SOIC	Small Outline IC
SOP	Small Outline Package
SOS	Silicon-On-Sapphire
SOT	Small Outline Transistor
SOW	Statement Of Work
SP	Signal Processor
SPDT	Single Pole Double Throw

SPICE	Simulation Program For Integrated Circuits Emphasis
SPS	Systolic Processing Superchip
SPST	Single Pole Single Throw
SRAM	Static RAM; Short Range Attack Missile
SREMP	Source Region Electromagnetic Pulse
SSI	Small Scale Integration
SSWS	Static Safe Work Station
STL	Stripline

TAB	Tape Automated Bonding
TC	Thermocompression (Bonding)
TCC	Temperature Coefficient Of Capacitance
TCE	Temperature Coefficient Of Expansion
TCR	Temperature Coefficient Of Resistance
TFE	Tetrafluoroethylene
Tg	Glass Transition Temperature
TGA	Thermal Gravimetric Analysis
TMA	Thermomechanical Analysis
TS	Thermosonic (Bonding)
TTL	Transistor Transistor Logic

UL	Underwriters Laboratory
US	Ultrasonic (Bonding)
ULSI	Ultra Large Scale Integration
UV	Ultraviolet

V

V	Volt
VCD	Variable Center Distance
VHF	Very High Frequency
VHSIC	Very High Speed Integrated Circuits
VLSI	Very Large Scale Integration
VPS	Vapor Phase Soldering

W

W	Watt
WIP	Work In Progress
WSI	Wafer Scale Integration

X

X	Reactance
X-MOS	High Speed MOS

Y

YAG	Yttrium Aluminum Garnet

Z

Z	Impedance
ZIP	Zigzag In Line Package
ZIF	Zero Insertion Force